Food and Health

Series Editors

The goal of this series is to provide coverage of emerging topics in food and health, using an interdisciplinary approach that considers health not only in a functional and human sense, but also in terms of external factors such as the environment. Titles in the series will address growing concerns about the future health, sustainability and quality of the food supply, as well as diet, and provide a home for books focusing on social and environmental concerns related to food.

Diane Mayerfeld
Editor

Our Carbon Hoofprint

The Complex Relationship Between Meat and Climate

Springer

Editor
Diane Mayerfeld
Division of Extension, Agriculture Institute
University of Wisconsin
Madison, WI, USA

ISSN 2509-6389 ISSN 2509-6397 (electronic)
Food and Health
ISBN 978-3-031-09022-6 ISBN 978-3-031-09023-3 (eBook)
https://doi.org/10.1007/978-3-031-09023-3

This Springer imprint is published by the registered company Springer Nature Switzerland AG
The registered company address is: Gewerbestrasse 11, 6330 Cham, Switzerland

Preface

I think of myself as an environmentalist. I have worked on environmental issues for more than 40 years, from marine and terrestrial habitat preservation to groundwater protection to sustainable agriculture. I also try to avoid causing excessive environmental harm through my personal lifestyle, and my food choices are an important part of that effort.

However, while I care deeply about the environment, I am not a fanatic. I commute to work and run in-town errands by bicycle, but my family does own a car, which we use more than absolutely necessary. When there is not a pandemic, we also take occasional plane trips, and though our home is energy efficient and low-waste by American standards, it could be smaller and even more efficient. Our food choices likewise compromise between environmental, economic, and social concerns and pleasure.

Since 1996 my work has focused on making agriculture more sustainable – on trying to reduce the pollution and habitat destruction so much farming causes, while also improving the way our agriculture and food system treats people. My knowledge of agriculture in the American Midwest led me to the conclusion that livestock are a critical part of a sustainable agroecology, and that well-managed grass-based cattle and sheep are particularly beneficial to agroecosystems.

The release of the United Nations *Livestock's Long Shadow* report and several widely read articles about the lower emissions intensity of industrial livestock challenged those conclusions. I dove into the literature, first to try to justify my original convictions, and then increasingly to try to really understand the impact of different livestock systems and what we might do to reduce that impact. This book is the result of that years-long effort.

What I found is that it's complicated. Livestock production does result in significant greenhouse gas emissions, but it is also still a critical component of sustainable agroecosystems. Both industrial and grass-based agriculture offer important ways to reduce emissions, and both systems have much room for improvement. Many on all sides of the discussion are guilty of exaggeration and of omission, and those distortions have prevented real dialogue and positive action.

For example, I just watched a slick opinion video called "Meet the People Getting Paid to Kill Our Planet," and it reminded me forcibly of why I wrote this book. The video criticizes the fact that there has been very little movement to regulate greenhouse gas emissions from agriculture in the United States and blames agricultural lobbying organizations for that lack of action.

Here's the thing – I agree with most of what the video says, including the message that American agriculture needs to do much more to address climate change. But at the same time I am infuriated by the deliberate distortions in the video. For example, they lead by saying that agriculture accounts for one third of global greenhouse gas emissions and that in the United States agriculture doesn't get enough of the blame for climate change – too much focus is on big oil and electricity generation. The implication that agriculture is the largest U.S. greenhouse gas emitter is far from the truth. In the United States, agriculture accounts for about a tenth of the country's greenhouse gas emissions, while electricity generation produces 30% and transportation 26 to 28% of greenhouse gas emissions. Other examples of misleading statistics and manipulative videography occur throughout the video.

Agricultural interests also frequently engage in distortion and misdirection when discussing climate change. For example, proponents of industrial livestock often omit greenhouse gases from feed production when they calculate livestock emissions, while proponents of grazing imply that the impressive soil carbon gains achieved by a tiny number of farmers can be achieved everywhere, without reductions in total meat production.

These extreme positions may fire up the base, but they just infuriate those they demonize and cause them to dig in their heels. They also cause complacency in the audiences they favor. My hope is that by finding the valid points on all sides of this discussion, and by challenging the distortions, this book can help move all of us to a respectful and honest examination of how to reduce livestock's carbon hoofprint. This effort is not about finding compromise in the middle, but rather about facing the fact that all of us need to think critically and act more effectively to address the real climate impacts of our food system. Only then can we move to effective action both by individuals and policy makers.

Many people helped with this book. Bob van Oort, Robbie Andrew, Nina Holmelin, Richard Teague, and Steve Apfelbaum contributed vital chapters. Jude Capper stepped in at the last moment to help write a chapter when another author didn't come through. Colleagues in Extension and at UW-Madison read early drafts, sent articles, and helped refine my thinking. Farmers and sustainable agriculture advocates in Wisconsin and beyond shared their experiences and insights. And friends and family not only tolerated but also supported my obsession with the topic, from my vegan brother to my meat-loving husband and children. Thank you all!

Madison, WI, USA Diane Mayerfeld

Contents

Chapter 1
How We Got Here, and Where We Need to Go: The Bitter Fight About Meat and Climate

Diane Mayerfeld

Abstract How much does animal agriculture contribute to climate change, and what is the best strategy to mitigate livestock's greenhouse gas emissions? This book examines three competing answers to those questions: the vegetarian view that livestock are the leading cause of climate change, the argument from industrial agriculture that modern production is so efficient that livestock emissions are negligible, and the view that grazing animals can actually help reduce climate change.

The debate among these positions is part of a broader discussion and negotiation around how to manage our global climate commons. A commons management framework underscores the need to base management and policy on an accurate understanding of the ecological processes involved in livestock's carbon hoofprint, as well as the need for honest negotiation among competing interests, and fair treatment of all participants using the resource.

Keywords Climate change · Commons · Diet · Greenhouse gas emissions · Livestock · *Livestock's Long Shadow*

A few years ago I was standing at the bus stop on a snowy morning, waiting to take the bus to work. My neighbor came up to wait for the same bus, and after we had exchanged greetings he asked "What is the least bad meat to eat for the climate?" Over the course of the conversation that ensued it became clear that he was looking for a simple answer. But I didn't have a simple answer to give him, even though (or perhaps because) I had been reading and thinking about the sustainability of meat for many years. Fortunately for my neighbor I had to get off the bus and head to my office before I talked his ear off. I know my co-authors have had similar encounters. This book is our attempt at a full response to Nick's question, and to many related conversations with friends, relatives, and colleagues.

D. Mayerfeld (✉)
Division of Extension, Agriculture Institute, University of Wisconsin, Madison, WI, USA
e-mail: dbmayerfeld@wisc.edu

D. Mayerfeld (ed.), *Our Carbon Hoofprint*, Food and Health,
https://doi.org/10.1007/978-3-031-09023-3_1

In 2006, the UN Food and Agriculture Organization released *Livestock's Long Shadow*, a report on the global environmental impact of livestock. One item from the executive summary captured the attention of the press and the public: the statement that livestock production emits 18% of all human caused greenhouse gases, making it the economic sector with the highest emissions.[1] (According to that report, transportation was the next biggest human source of greenhouse gas emissions, accounting for 17% of the total.) Even though *Livestock's Long Shadow* came out more than 15 years ago, that one statement from the executive summary still strongly influences both academic and public discourse around the impact of animal agriculture.

Although a significant fraction of the U.S. population remains skeptical about the role of humans in climate change, the issue has great emotional resonance.[2] *Livestock's Long Shadow* dedicated about 200 pages to livestock's effects on water supply and quality and on biodiversity and only about 50 pages to its air quality and climate impacts, but the climate discussion has had a far greater public impact than the rest of the report. Comments about climate impacts of meat regularly crop up in the news and popular media. However, there is a great deal of debate about what the best response is for people and the climate.

Debate drives science and can build knowledge, but much of the discussion around the role of livestock in climate change has had the perverse result of entrenching groups in their views – and giving them a false sense that *they* don't need to change. In the United States and western Europe three viewpoints have dominated the discussion. The **vegetarian** viewpoint embraced by many individuals and groups who were already opposed to raising animals for meat is that the findings in *Livestock's Long Shadow* and related publications prove that meat production and consumption is environmentally ruinous and that eating less meat and working toward the elimination of animal agriculture are the most important actions people can take to reduce the severity of climate change. The **efficiency** viewpoint promoted by groups and individuals working in the mainstream meat and dairy sectors in the U.S. and Europe is that modern, intensive livestock production is so efficient that their greenhouse gas emissions are negligible. The **soil carbon** viewpoint promoted by advocates of intensively managed grazing is that good grazing management can sequester significant amounts of carbon in the soil and so can actually reduce the total amount of greenhouse gases in our atmosphere. Each of these three viewpoints contains some truth, but also some serious omissions that can lead to false complacency.

[1] Steinfeld et al. (2006).

[2] Agreement that climate change is occurring and is caused by humans is increasing in the U.S., but a significant part of the public is still unconvinced. In 2020 60% of Americans agreed that climate change is human-caused, leaving 40% who either do not agree the climate is changing or believe the causes are primarily natural. In rural areas there is even more hesitation to attribute climate change to human activity; in the 2020 Iowa Farm and Rural Life Poll 81% of respondents agreed that climate change is occurring, but only 18% thought the changes are due primarily to human activities, while 23% attributed climate change to natural causes, and 40% said it was due equally to human and natural causes (Arbuckle, 2020; Ballew et al., 2019; Wang et al., 2021).

Chapters 2, 3, 4, 5, 6 and 7 of this book will dive into the details and limitations of each viewpoint, but this introduction will look at the broader context of the debate and try to address the question of how these three very different interpretations arose and why it is so important to move beyond these fragmented and polarized views.

1.1 A Challenge of the Commons

One way to think about climate change is as a wicked problem of commons management. Commons management and wicked problem are both complex social science concepts. Bear with me as I briefly introduce both concepts and their relevance to the debate around livestock and climate. All environmental problems have elements of both common resource management and "wickedness", but climate change has both challenges to an extreme degree. These challenges help drive broad policy conflicts around climate, as well as the specific arguments over the role of livestock in causing climate change that are the focus of this book.

In Garret Hardin's 1968 essay "The Tragedy of the Commons" he recounts a fable of a community owned pasture – a commons – that is ruined when each (or even one) member of the community calculates that they will benefit by putting more cows on the pasture. Hardin was not particularly interested in the grazing of cows. He used the idea of a community pasture as a metaphor for all natural resources held in common. Still, the metaphor of a common pasture is particularly appropriate for a discussion of livestock's impact on climate! Although there are many situations where commonly held land is managed well, there is some truth to the idea that resources held in common can be difficult to protect. That difficulty tends to grow as the size of the group managing the resource grows, especially when the people involved have no direct social ties with each other.[3]

Hardin proposed two ways to protect natural resources: by privatizing them or by subjecting them to strict government regulation. Private ownership is no guarantee of good resource management, as our continuing battle with privately owned eroding agricultural fields and hazardous waste sites attests. But even if private ownership could guarantee good stewardship, there are many natural resources that are simply not suitable for privatization because they don't lend themselves to excluding or controlling access.[4]

---

[3] Feeny et al. (1990), Hardin (1968), Ostrom et al. (2002).

[4] Feeny et al. (1990), Gibson et al. (2000). Along with excludability, scholars of property and commons management talk about the subtractability or rivalry of goods. That is, if the availability and quality of a resource is not affected when people use it, then there is no reason to manage it as private property. Once upon a time humans treated clean air and the climate as non-subtractable goods, but as both population and consumption have grown it has become clear that they are subject to subtractability, at least through abuse. It is hard to think of a non-rival material good with current population and consumption levels.

What does it mean that some resources are not excludable (as commons scholars would say), and why does that matter? Let's look at two examples: land and air. You can exclude or control access to land by building a wall around it. In a society that regards land as a private resource you probably don't even have to build a wall or fence. People recognize property boundaries based on physical or cultural features on the land, and they stay away from land owned by others because they respect property rights or because they fear being prosecuted for trespass, or for both reasons. So land is an excludable resource. But how do you exclude people's access to air? Air is mobile and invisible and everywhere. Perhaps one could figure out a way to turn air into a possession, but it isn't easy or obvious. Also, our lives depend on constant access to clean air, so the idea of privatizing air raises huge ethical questions.[5]

Climate is a non-excludable resource. With every breath and every action each of us depends on and affects the balance of gases that regulate our climate, if only a little bit. And not just humans, but every living organism. So privatizing climate is not a solution for our climate crisis.

What about the other solution Hardin proposed, government regulation? Indeed, government regulation will need to be part of how we rein in climate change, but dealing with greenhouse gas emissions is still a problem of the commons. Although it is not usually described that way, government regulation is really just a special case of commons management. Like any community that tries to manage a common property, the state must negotiate issues of fairness and enforcement and must also cope with changes in the resource or the context that surrounds it. And as with non-formal communities that seek to manage common resources, governments sometimes fail to adequately protect resources when powerful interests stand to benefit from exploiting those common resources.[6]

Moreover, we don't have a single world government, so all nations will need to participate in regulating greenhouse gas emissions. So we have two levels of climate commons negotiations: the global negotiations about how many emissions each country gets to have, and the negotiations within each nation about how to reach those emissions targets.[7]

Scholars have identified many factors that make common resource management more likely to succeed. Three key factors that contribute to successful commons management are central to the purpose of this book. They are:

1. understanding of the ecological processes involved,
2. honest negotiation, and

---

[5] The feudal and capitalist paradigms of private land ownership also raise ethical questions, but those are far beyond the scope of this book.

[6] This lack of enforcement is sometimes termed regulatory capture. Powerful interests can also prevent governments from even enacting laws to protect resources, or significantly delay such laws. (Dillon et al., 2018; Shimshack & Ward, 2005)

[7] In fact, there are commons negotiations around climate issues at many levels within a nation: at the state or provincial level, within trade associations, within businesses, and even within individual farms and households.

3. fair treatment of all participants using the resource.

Clearly, an understanding of the ecological factors affecting the common resource is critical to successful management. And the answers are not always simple. Consider again the example of a common pasture that is degrading. In Hardin's fable he assumes the cause of the degradation is that there are too many cows eating the grass. It is true that when the number of cows exceeds the carrying capacity of a pasture they will eventually kill the grass by overgrazing. But there are many other factors that might lead to a degraded pasture. In our hypothetical pasture the problem may not be that there are too many animals for the land that is available. Instead, perhaps there is just one water source, and so the cows have severely overgrazed the area near the spring, while the parts of the pasture that are far away from water have become overgrown and shrubby. Reducing the total number of cows on the pasture would allow the grass a little distance from the spring to recover, but the area near the spring would still be overgrazed, the edges of the pasture would be even more overgrown, and the community would have less milk and meat. A better solution would be to develop other water sources and rotate the animals around all the common land. This approach would allow each portion of the pasture to recover and regrow, keep the majority of the common land in grass, and provide more milk and meat to the local population.

Our challenge is that we don't have a full understanding of the ecological factors involved in climate change, but we cannot afford to wait for action until we do. Wait, you say, don't we have a good understanding of the drivers of climate change? At some levels we do. There is overwhelming scientific consensus that human activity is causing climate change, and that the biggest factor is the burning of fossil fuels – coal, petroleum, and natural gas. Fossil fuel combustion takes the carbon that plants pulled out of the atmosphere and stored in the ground over millions of years and dumps it back in minutes into the atmosphere as carbon dioxide. We also know that some kinds of land use change, especially deforestation, result in significant greenhouse gas emissions. But when we get to the details of trade-offs between similar activities the picture often gets muddier, and careful studies and models can point to different best practices, as Chaps. 4 and 6 will show. And certain kinds of emissions, such as nitrous oxide (N_2O) emissions from soils, are difficult and expensive to monitor. At the same time, given the political success of climate change denial in the U.S., scientists and activists are cautious about discussing uncertainties about specific greenhouse gas emission sources.

This uncertainty brings us to the second factor, that of honest negotiation. It is a common negotiation tactic to mention only the information that favors your side. When you are selling an item it is standard practice to emphasize its merits and downplay or ignore its faults. But when the consequences of incomplete information are very grave, omitting negative information is no longer an acceptable strategy. Thus, to protect the health and life of the buyer, we have laws requiring a person selling their home to disclose if it has asbestos, and requiring companies selling food products to list all the ingredients in that food on the label. The consequences of unchecked climate change threaten the well-being and lives of billions of people, as well as other species. Yet our negotiations around climate change, both among economic sectors and among countries, are filled with deliberate omissions. Our incomplete knowledge and uncertainty might make this approach to negotiation seem acceptable – after all there is uncertainty about the greenhouse gas emissions we are ignoring. Moreover, the other side is also using data selectively. But we must not use uncertainty or incomplete knowledge or the questionable tactics of opposing viewpoints as an excuse for omitting critical information about the causes of climate change. As a planet we cannot afford to negotiate our responsibility to reduce greenhouse gas emissions without full honesty.

The third factor that encourages successful management of common resources, fair treatment of all participants, brings us to the wickedness of the climate challenge. The term "wicked problem" refers to complex social policy problems that don't have widely accepted technical or consensus solutions. That is not to say that technology has no part to play in addressing wicked problems, but rather that different technical solutions have unequal costs and benefits, which leads to fights over who should bear the costs of the technology and which technology should be chosen. Those fights can prevent all sides from moving ahead with solutions.

With climate change we are surrounded by technologies that can greatly reduce our greenhouse gas emissions. Here are just a few examples:

- We have the technology to generate electricity from sunshine, wind, hydropower, and other renewable sources without burning fossil fuels, but that would place financial burdens on utilities that invested in coal and natural gas generation, and on companies that mine, transport, and process coal and natural gas, as well as on coal miners and their families. So the energy sector resists changing our energy sources.
- We have the technology to build efficient public transportation powered by renewable energy that could replace much of our plane travel and private car use, but that would require government investment and therefore higher taxes. It would also hurt the profits of airlines, car companies, oil companies, and all the people who work for them. So most countries continue to reduce their investment in efficient public transit, and at least in the United States auto manufacturers continue to build and sell gas-guzzling trucks and cars.[8]
- We have the technology to build more energy efficient homes and offices, but that would raise construction costs and reduce energy company profits, so most new construction pays little attention to energy efficiency or the greenhouse gas implications of construction materials such as concrete.
- We have the technology to reduce food waste and consume less stuff (clothes, electronics, etc.) but we worry that would reduce the profits of those who produce and process and sell that food and stuff, from farmers to factories to super-

[8] U.S. Environmental Protection Agency (2021). Recently some auto manufacturers have expressed an intention to switch to electric vehicles (Bomey, 2021; MacDuffie & Light, 2021); how well they follow through will undoubtedly depend on the policy and market environment of the next few decades.

stores. And all of these options could increase costs for some consumers, at least temporarily. So we continue to overproduce and waste food and consumer goods.
- We have the technology to limit how many children a person bears, and stabilizing human population will sooner or later be a critical element of reducing or reversing our greenhouse gas emissions. But while that technology has allowed some of us to choose how many children to have and when to have them, limits on birth have also at times been imposed on individuals and groups without their consent, so that technology is particularly fraught with tensions around the power of different groups.[9] Allowing women to regulate their own fertility can also improve the economic and social power of women and thus threaten traditional male hierarchy.

These are just a few examples of technologies we could use to manage greenhouse gas emissions. The problem is that adopting any of these technologies would impose burdens or reduce privilege for at least some people, and those costs will probably not be shared equally. As many have pointed out, there would also be sectors and people who would benefit from the adoption of climate-friendly technologies. However, that point is not likely to persuade those who fear to suffer from that technological change, many of whom have considerable political power. In fact, knowing that someone else may profit from your loss could make you oppose the change even more.

And so, much of the debate around meat and climate change is at its core less an argument about the facts of climate change than a negotiation about who needs to change, and who needs to pay for change. And unfortunately, so far in most cases that negotiation has followed the lines of bargaining for advantage, where each sector has selectively used information (and in some cases disinformation) to minimize its responsibility for addressing climate change.

The temptation to negotiate with selective information is strengthened by the fact that climate change is very hard to wrap our minds around. Greenhouse gases are invisible and odorless, and they come from a very wide variety of sources. Moreover, there is no direct connection between any specific emission of a greenhouse gas and a change in the weather.[10] When a factory spills a pollutant into a river there is a direct and often visible or smell-able link between that action and the resulting degradation of water quality. Pollutants discharged into the ground from underground tanks or leaking landfills or other sources may not be visible at the surface, but with a reasonable understanding of hydrogeology we can usually directly connect water contamination to a specific source or cluster of sources. Sometimes pollution problems come from many small sources – like smog from hundreds of thousands of cars, or fertilizer runoff from millions of acres. This type of pollution (called non-point source pollution) is harder to see and trace, and therefore harder for governments to regulate. But although the link is not visible, the cars and smog are

[9] Bell et al. (2021), Patel (2017).

[10] I use the term weather intentionally. Although weather is not the same thing as climate, we experience climate change in the form of weather.

connected geographically, and it makes logical sense that many tiny sources can add up to a big problem.

With climate change that geographic connection is gone. Some of the greatest climate change impacts to date are in the Arctic, an area that is sparsely populated and not historically a major source of greenhouse gas emissions. The link between a car trip in the Midwestern U.S. and the loss of glaciers in Greenland or flooding in Bangladesh seems abstract, to say the least. Although science tells us that driving cars releases carbon dioxide that heats the planet and causes both sea level rise and more intense storms, that connection is neither visible nor intuitively logical.

Yes, managing resources in common is difficult, and climate is a particularly challenging resource for commons management (more about that below). But that is no reason to give up on commons management. *At some level all resources are held in common*. For example, land is seen as a resource that can be held either as private property or as a commons. But even privately held land is a resource managed in common over time. In this context Chief Seattle got it mostly right when he said "We do not inherit the Earth from our ancestors—we borrow it from our children." Actually, land, water, air, biodiversity, and the Earth are resources shared over time with our ancestors, our neighbors, our children, and with their children and grandchildren and great-grandchildren and so on (or at least so I hope).

In a negotiation about an actual pasture managed in common we can probably assume that the people using the common land have social and political ties. They may or may not be friends, but they live in the same community and likely speak the same language and are part of the same political structure. But greenhouse gas emissions come from around the globe, and so humans face the unprecedented and "wicked" challenge of collaborating across wide cultural, political, geographic, and economic divides to regulate the emissions that threaten our climate.

Elinor Ostrom notes that internal factors that can hinder a group's ability to successfully manage a common resource include an inability to communicate with each other, lack of opportunity to develop trust, and no sense that they share a common future.[11] Modern technology has done much to overcome communication barriers, but language and cultural differences, as well as sheer numbers inhibit humanity's ability to develop trust. Climate change will affect all of us, but some more than others, at least in the next 50 years. Ostrom also notes that when powerful groups or individuals gain from the abuse of the resource while less powerful groups suffer, this disparity in power can prevent good resource management.

In the classic tragedy of the commons fable the cause of degradation is clear – too many cows on too little land eat and trample all the grass. With greenhouse gas emissions we can't see the links between our human actions and our climate's reaction; instead we are asked to trust complicated scientific models to tell us which of our daily activities will doom future generations. This invisibility makes it easy and tempting to absolve ourselves and blame someone else for the problem.

[11] Gibson et al. (2000), Ostrom et al. (1999).

Also, many different activities emit greenhouse gases. To belabor the metaphor of the commons even further, imagine the town meeting if the residents use the commons for playing soccer and for parking as well as for grazing. And rather than just cows, some people are putting sheep on the common, while others are pasturing poultry, and a few let their horses graze there. Every one of those activities puts pressure on the grass, but with all of them happening on the same piece of land it is much harder to figure out how to fairly regulate each use. Chances are that the soccer players will say the solution is to reduce the livestock pressure. The farmers will argue with each other about whether one horse is a bigger problem than six sheep, except when they unite to propose that banning soccer and parking is clearly the first step to restoring the grass. And the drivers will suggest paving the site to solve the mud problem.

So it has been with the finding that livestock production is a major source of greenhouse gas emissions. Many people who were critical of meat consumption to begin with have seized on the *Livestock's Long Shadow* report to advocate reducing or even eliminating meat and dairy in our food system. People working in conventional livestock production argue that intensive modern animal rearing is so efficient that its greenhouse gas emissions are negligible. And those working in grass-based agriculture maintain that, just as the North American prairies, Asian steppes, and African savannas stored carbon when vast herds of bison, antelope, gazelles, and other wild herbivores roamed, so modern grazing techniques can help pull CO_2 out of the atmosphere and store it in the soil. There is some truth to all three of these positions, but each also has major problems, as we will explore in the following chapters.

Several other factors fuel the discussion: a long history of concern about eating animals, the idea that diet change is an action individuals can take without the need for difficult structural or political change, and economic stress and fear in the livestock sector.

The food we eat has always had meaning beyond nutrition. Our food choices are shaped by culture and fashion as well as our bodies' physical demands, and we use the food we consume to communicate status, identity, and values. We also judge those around us by the food they eat and how they eat it, from potential romantic partners at a date to job candidates at a lunch or dinner interview. For the interview meal multiple websites advise the interviewee about details of formal etiquette, from which piece of tableware to use to what types of food to order and which to avoid. In other settings food judgments can run in the opposite direction; a politician who declines working class food or eats it incorrectly (think New York mayor Bill deBlasio eating pizza with a knife and fork)[12] invites ridicule and contempt. Thus, the debate around the role of livestock in climate change takes place in context of a culture predisposed to tell people what they should eat.

In particular, people have long felt deep ambivalence about eating meat. For thousands of years many of the world's major religions have wrestled with the

[12] Grynbaum (2014).

ethics of killing animals for food. Judaism and Islam prohibit the eating of pork and impose a set of special rules and rituals on slaughtering any livestock for meat. Hinduism prohibits the slaughter and eating of beef, and many devout Hindus abstain from eating any meat. Although most forms of Christianity do not prohibit meat, giving up meat, temporarily or permanently (for Lent or on Fridays, for example), is seen as a virtuous act. Many other religions and spiritual traditions grapple with meat consumption, and even without explicit religious prohibitions many cultures limit the types of animals that are acceptable to eat.

At the same time, many religious and secular traditions celebrate meat. In the Judeo-Christian story of Cain and Abel, God favors the sacrifice of lamb over that of grain, and the Old Testament is filled with instructions for animal offerings. Animal sacrifice and feasting are central to ancestor veneration in sub-Saharan Africa, to Islam at Eid al-Adha, to the Jewish Passover holiday, and doubtless to many other religious traditions. Meat also holds center stage in secular feasts, such as the Thanksgiving turkey.

But only relatively recently has eating meat become a secular environmental concern. In 1971 Frances Moore Lappé wrote *Diet for a Small Planet.*[13] In that book she pointed out that modern meat production in North America and Europe relies on feeding livestock grains that humans could eat, and that we lose a lot of food energy in the process. As a result, raising meat uses much more land per calorie and per gram of protein than growing grains, beans, lentils, and other high-protein seeds. The book's main argument was that if more people ate less meat there would be enough food for all, and people in both the U.S. and around the world would no longer need to go hungry. The original *Diet for a Small Planet* did not touch on climate change. In those days the greenhouse effect, as climate change was called then, was attributed purely to the use of fossil fuels, and fears about air and water pollution and about running out of food due to population growth loomed much larger in the public consciousness. But Lappé did touch briefly on soil degradation, the amount of land dedicated to growing feed for livestock, the amount of water needed to produce meat, and the use of pesticides.

This small book changed the way many thought about food. Even people who did not read the book or change their diet became aware that their food choices could have environmental and social consequences beyond their personal health and budget. Today, statements about the social and environmental impacts of food abound, from Michael Pollan's books to academic articles to advertising copy and ecolabels. The idea that the food we eat has implications for our environment and society crops up in the food section of major newspapers, as well as academic journals, and of course in food blogs and websites and social media.

In his book *In Defense of Food* Michael Pollan points out that most contemporary Americans have lost the clear consensus and guidance about what to eat that traditional food cultures offered. The loss of a clear food culture is increasingly true for European cultures and other wealthy nations. At the same time that we are

[13] Lappé (1971).

offered a dizzying array of foods, we are subject to a barrage of conflicting instructions about what we should and should not eat. From governmental nutritional guidelines to instruction in elementary and secondary school to restrictions on what types of food people can buy with government food aid to advice from our employer or health insurance carrier, we are told in detail what we should and should not eat. Most of those instructions are based on ideas about nutrition and individual health, but I believe they help set the stage for the notion that it is acceptable, even virtuous, to tell people what to eat.

A magazine article about Gidon Eshel, one of the leading scholars documenting the environmental and climate impacts of meat in our diets, quotes one of his students who was raised as a vegetarian for cultural reasons "I've become aware of the immense nutritional and environmental benefits of a plant-based diet. I love being able to advocate for my lifelong diet with incontrovertible scientific evidence."[14] Livestock farmers have much the same response when they hear about a study that documents the climate benefits of modern livestock production or of grazing. We all engage in confirmation bias. It is a necessary strategy in a world that contains far more information than the human brain can process. But we need to resist it when it threatens our planet.

1.2 Livestock's Long Shadow

Before moving to the next chapter, it is worthwhile to review briefly how the seminal *Livestock's Long Shadow* report reached its conclusions about the greenhouse gas contributions of livestock production. Figure 1.1 summarizes the report's estimates of the proportion of annual greenhouse gas emissions due to different aspects of livestock production. These estimates were for the early 2000s, mostly based on emission reports and livestock production figures and trends from the 1990s. The Food and Agriculture Organization has since significantly revised its calculations of livestock's greenhouse gas emissions,[15] but the figures from the original report still influence the debate and underpin the argument to greatly reduce consumption of meat and dairy.

These estimates are based on highly incomplete knowledge, so while the orders of magnitude are probably right, the precise numbers are almost certainly not. Nevertheless, it is clear that three aspects of livestock management stand out in the magnitude of their emissions: land use change, enteric fermentation, and manure. Let's look a little more at each of these emissions pathways and what the report said about them.

[14] Estrada (2017).

[15] Food and Agriculture Organization of the United Nations (2016), Gerber (2013), Ritchie (2020). Although livestock emissions continue to rise, global emissions from transportation and electricity generation are rising slightly faster, so the relative contribution of livestock agriculture on a global scale is gradually declining, even as absolute emissions increase.

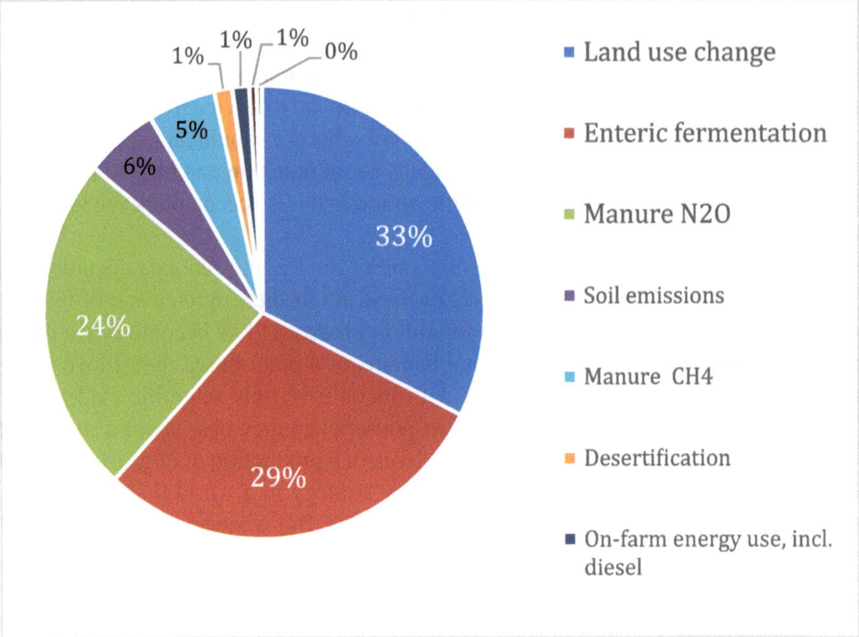

Fig. 1.1 Relative contributions of global livestock greenhouse gas emission sources identified in *Livestock's Long Shadow*

Land use change can be good or bad for the climate. Plants use photosynthesis to convert carbon dioxide from the air and water to carbohydrates. Those carbohydrates provide the energy plants need to live, and the building blocks for plant growth, as well as food for animals and microorganisms. In healthy ecosystems the plants pull more carbon out of the atmosphere than they, and the animals and microorganisms that consume them, need. That extra carbon is stored in living biomass like tree trunks and soil bacteria and fungi, and as carbon compounds in the soil. (Over millions of years that extra carbon can turn into oil, coal, and natural gas, given the right conditions.) But when actions like deforestation or plowing severely disturb a plant community, the remaining plants cannot photosynthesize enough to feed themselves, plus all the animals and microorganisms that depend on them. In those conditions microorganisms consume carbon that has been stored in the soil and in plants and animals, and respire that stored carbon back to atmosphere as CO_2. If the original ecosystem was a forest, much of the carbon stored in the trees may also be converted to CO_2 through burning. If timber is removed for commercial use the carbon in that wood may continue to be stored, but in any forest much of the woody biomass is not marketable and so is burned or allowed to decompose.

Land use change from natural ecosystems like forests to heavily disturbed ecosystems like annual agriculture releases large amounts of CO_2 over a period of months. Land use change in the other direction, say from annual agriculture to a forest, takes carbon out of the air, but it does so very slowly, over decades or centuries. *Livestock's Long Shadow* estimated that at least half of the deforestation occurring in Brazil and Bolivia in the early 2000s resulted in either pasture for cattle or cropland producing livestock feed (especially soybeans) and so attributed those CO_2 emissions to livestock. Worldwide, at the end of the twentieth century and beginning of the twenty-first most deforestation is occurring in the Amazon basin, sub-Saharan Africa, and southeast Asia, while forest land is stable or even growing slightly in Europe, North America, northern Asia, and other parts of Latin America. European livestock sectors can contribute indirectly to land use change emissions when they rely on livestock feed imported from land subject to recent clearing, but the U.S. produces far more livestock feed than it uses and so does not import feed from countries with significant deforestation. Thus, the very significant emissions attributed to livestock from deforestation depend heavily on how and where the meat is raised.

Enteric fermentation is the process where microbes in the digestive tract break down plant parts (such as cellulose in grass) that the animal cannot digest on its own and produce byproducts that the animal can use. Although many animals, including humans, use microbes to assist with digestion, ruminants have digestive systems that are specially adapted to maximize this microbial symbiosis. This mutually beneficial relationship with microorganisms allows cattle, sheep, and other ruminants to thrive on grasses and other forages that are high in cellulose and low in simple sugars. However, in addition to breaking down plant fibers and producing volatile fatty acids and amino acids that the animal can absorb, the bacteria and fungi in the rumen also produce carbon dioxide and hydrogen (CO_2 and H_2) and other byproducts. Another group of microorganisms called methanogens lives on these byproducts, in turn producing methane (CH_4) and water as waste products. The ruminants emit this methane both as they breathe out and belch, and less often through the back end of the digestive tract.[16] Methane is a potent greenhouse gas.

The number of methanogens and the amount of methane emitted by ruminants varies depending on the animal's diet and individual genetics. In general, methane emissions are thought to be lower if the animal's diet is lower in fiber and higher in rapidly degradable carbohydrates. This means that cattle on high quality pasture likely emit less methane than cattle grazing over-mature or woody forages, and cattle fed grain are expected to have even lower methane emissions from enteric fermentation.[17] *Livestock's Long Shadow* estimated that Asia (excluding Japan), Africa, and Central and South America accounted for about 76% of methane emissions from enteric fermentation, while North America, Europe, Oceania, and Japan accounted for 24%. And this large source of livestock greenhouse gas emissions is almost entirely associated with ruminants, especially cattle. Pork and poultry have negligible greenhouse gas emissions from enteric fermentation.

Finally, **manure** produces significant methane and nitrous oxide emissions. Unlike enteric fermentation, manure methane emissions are not limited to

[16] National Academies of Sciences, Engineering, and Medicine (2016).

[17] Danielsson et al. (2017), Dini et al. (2012), Gerber et al. (2013).

ruminants, and pigs and poultry accounted for about 53% of global methane emissions from manure. Methane emissions occur when methanogens feed on the nutrients in the manure under anaerobic (oxygen-poor) conditions. These anaerobic conditions happen when manure is stored in lagoons, a practice associated with industrial livestock production and concentrated in Europe and North America, but increasing in other parts of the world, especially in China. Thus, North America, Europe, Oceania, and Japan accounted for 53% of global methane emissions from manure, and China accounted for another 22% of global emissions.

Manure greenhouse gas emissions in the form of nitrous oxide (N_2O) were roughly five times greater than the methane emissions from manure. Nitrous oxide is an even more potent and long-lasting greenhouse gas than methane, and according to the report livestock manure accounted for 65% of global human-caused nitrous oxide emissions. The nitrogen cycle is complex, and many factors influence what happens to the nitrogen in manure and how much of it is converted to N_2O. Some of these factors are the diet of the animal and resulting nitrogen content of the manure, the concentration of the manure as it is applied to soils, soil temperature and moisture, and the ability of growing plants to take up and use nitrogen. As a result, the estimates of N_2O emissions are particularly uncertain. Nitrous oxide emissions from manure are not heavily influenced by whether the livestock are grazing or fed in confinement. Estimated manure N_2O emissions from Asia, Africa, and Central and South America were roughly 66% of the global total, with 34% of manure N_2O emissions coming from North America, Europe, Oceania, and Japan.

This book examines each of the three main viewpoints about livestock and climate in turn, with a focus on North America and Western Europe. As we have already seen, there are significant differences in greenhouse gas emissions from livestock production in different parts of the world, so it is important to understand the specific impacts for a region. In Chap. 2 Bob van Oort, Robbie Andrew, and Nina Bergan Holmelin lay out the arguments for reducing meat consumption in Europe in order to meet global climate goals, based on the case of Norway. Chapter 3 critiques some of the excesses and distortions of arguments that livestock agriculture is the main cause of climate change. Chapter 4 briefly summarizes how advances in livestock science have enabled highly efficient milk and meat production in the United States and Western Europe, with relatively low greenhouse gas emissions. Chapter 5 discusses the limits of efficient production for reducing livestock's climate impact. In Chap. 6 Steven Apfelbaum and Richard Teague describe the symbiotic relationship between grasslands and grazing animals and the role grazing plays in storing soil carbon, as well as supporting biodiversity and other ecological services. Chapter 7 looks at some of the limits of grazing and critiques some exaggerated claims about the climate benefits of grazing.

I should note that Chaps. 3, 5, and 7 are not refutations of the preceding chapters; rather they are critiques of ways that valid observations about greenhouse gas emissions from various types of livestock agriculture presented in those chapters have been exaggerated and distorted to capture attention and advance a particular interest.

Finally, Chaps. 8 and 9 make recommendations both for individual action and for a collective response to reduce our carbon hoofprint. Climate change is a challenge unlike any other that humankind has ever faced. In this drama of the commons we cannot afford to fail. While emphasizing the truth that supports one's position is a time-honored negotiating tactic, in this effort to manage our planet's climate we cannot work from distorted information. Nor can we let the real gaps in our knowledge keep us from action.

Bibliography

Arbuckle, J. G. (2020). *Iowa farm and rural life poll: 2020 summary report* (Extension report SOC 3094). Iowa State University Extension.

Ballew, M. T., Leiserowitz, A., Roser-Renouf, C., Rosenthal, S. A., Kotcher, J. E., Marlon, J. R., Lyon, E., Goldberg, M. H., & Maibach, E. W. (2019). Climate change in the American mind: Data, tools, and trends. *Environment: Science and Policy for Sustainable Development, 61*(3), 4–18. https://doi.org/10.1080/00139157.2019.1589300

Bell, M. M., Ashwood, L., Leslie, I. S., & Schlachter, L. H. (2021). *An invitation to Environmental sociology* (6th ed.). SAGE.

Bomey, N. (2021, March 11). Are electric vehicles poised to kill the gasoline engine car? Welcome to the "golden age" of EVs. *USA Today*. https://www.usatoday.com/story/money/cars/2021/03/11/electric-vehicles-tesla-gm-lucid-rivian-volvo-gas-cars/4581584001/

Danielsson, R., Dicksved, J., Sun, L., Gonda, H., Müller, B., Schnürer, A., & Bertilsson, J. (2017). Methane production in dairy cows correlates with rumen methanogenic and bacterial community structure. *Frontiers in Microbiology, 8*. https://doi.org/10.3389/fmicb.2017.00226

Dillon, L., Sellers, C., Underhill, V., Shapiro, N., Ohayon, J. L., Sullivan, M., Brown, P., Harrison, J., Wylie, S., & The "EPA Under Siege" Writing Group. (2018). The Environmental Protection Agency in the early trump administration: Prelude to regulatory capture. *American Journal of Public Health, 108*(S2), S89–S94. https://doi.org/10.2105/AJPH.2018.304360

Dini, Y., Gere, J., Briano, C., Manetti, M., Juliarena, P., Picasso, V., Gratton, R., & Astigarraga, L. (2012). Methane emission and milk production of dairy cows grazing pastures rich in legumes or rich in grasses in Uruguay. *Animals, 2*(2), 288–300. https://doi.org/10.3390/ani2020288

Estrada, I. (2017). *Rethinking the American diet*. Radcliffe, Winter.

Feeny, D., Berkes, F., McCay, B. J., & Acheson, J. M. (1990). The tragedy of the commons: Twenty-two years later. *Human Ecology, 18*(1), 1–19. https://doi.org/10.1007/BF00889070

Food and Agriculture Organization of the United Nations. (2016). *Environmental performance of large ruminant supply chains: Guidelines for assessment*. FAO. http://www.fao.org/3/i6494e/i6494e.pdf

Gerber, P. J. (Ed.). (2013). *Tackling climate change through livestock: A global assessment of emissions and mitigation opportunities*. Food and Agriculture Organization of the United Nations.

Gerber, P. J., Hristov, A. N., Henderson, B., Makkar, H., Oh, J., Lee, C., Meinen, R., Montes, F., Ott, T., Firkins, J., Rotz, A., Dell, C., Adesogan, A. T., Yang, W. Z., Tricarico, J. M., Kebreab, E., Waghorn, G., Dijkstra, J., & Oosting, S. (2013). Technical options for the mitigation of direct methane and nitrous oxide emissions from livestock: A review. *Animal, 7*(s2), 220–234. https://doi.org/10.1017/S1751731113000876

Gibson, C. C., McKean, M. A., & Ostrom, E. (2000). Explaining deforestation: The role of local institutions. In *People and forests: Communities, institutions, and governance* (pp. 1–26).

Grynbaum, M. M. (2014, January 10). A fork? De Blasio's way of eating pizza is mocked. *New York Times*. https://www.nytimes.com/2014/01/11/nyregion/de-blasio-skewered-for-eating-pizza-with-utensils.html

Hardin, G. (1968). The Tragedy of the Commons. *Science, 162*(3859), 1243–1248.

Lappé, F. M. (1971). *Diet for a small planet*. Friends of the Earth/Ballantine.

MacDuffie, J. P., & Light, S. E. (2021, March 15). EV turning point: Momentum builds for U.S. electric vehicle transition. *Yale Environment, 360*. https://e360.yale.edu/features/ev-turning-point-momentum-builds-for-u.s.-electric-vehicle-transition

National Academies of Sciences, Engineering, and Medicine. (2016). *Nutrient requirements of beef cattle* (8th ed.). National Academies Press.

Ostrom, E., Burger, J., Field, C. B., Norgaard, R. B., & Policansky, D. (1999). Revisiting the commons: Local lessons. *Global Challenges, 284*, 6.

Ostrom, E. E., Dietz, T. E., Dolšak, N. E., Stern, P. C., Stonich, S. E., & Weber, E. U. (Eds.). (2002). *The drama of the commons*. National Academy Press.

Patel, P. (2017). Forced sterilization of women as discrimination. *Public Health Reviews, 38*(1), 15. https://doi.org/10.1186/s40985-017-0060-9

Ritchie, H. (2020, September 18). Sector by sector: Where do global greenhouse gas emissions come from? *Our World in Data*. https://ourworldindata.org/ghg-emissions-by-sector

Shimshack, J. P., & Ward, M. B. (2005). Regulator reputation, enforcement, and environmental compliance. *Journal of Environmental Economics and Management, 50*(3), 519–540. https://doi.org/10.1016/j.jeem.2005.02.002

Steinfeld, H., Gerber, P., Wassenaar, T. D., Castel, V., Rosales, M. M., & de Haan, C. (2006). *Livestock's long shadow: Environmental issues and options*. Food and Agriculture Organization of the United Nations.

U.S. Environmental Protection Agency. (2021). *The 2020 EPA automotive trends report: Greenhouse gas emissions, fuel economy, and technology since 1975* (EPA-420-R-21-003). U.S. Environmental Protection Agency. https://www.epa.gov/automotive-trends/download-automotive-trends-report

Wang, X., Leiserowitz, A., & Marlon, J. (2021, March 31). *Explore Climate Change in the American Mind*. Yale Program on Climate Change Communication. https://climatecommunication.yale.edu/visualizations-data/americans-climate-views/

Chapter 2
The Consequences for Climate of Meat Consumption

Bob van Oort, Robbie Andrew, and Nina Bergan Holmelin

Abstract Livestock production generates a significant share of total global greenhouse gas emissions and poses other, non-climatic challenges to society and nature. At the same time, livestock and meat also provide social and environmental benefits. Acknowledging the broader importance of livestock, this chapter focuses on meat production and consumption's impact on climate, with an emphasis on the role of ruminants. Regional differences in production and consumption, the role of metrics in comparing different climate gases, and effects of grazing on soil carbon and albedo are discussed both in a general global context, and in more detail in a case study for Norway. The chapter concludes that there remains a role for livestock in the future, but that improvements in production and reductions in ruminant meat consumption, combined with targeted measures in other sectors, are key ingredients to limit future warming beyond global targets.

Keywords Beef · Climate change · Consumption · Livestock · Meat · Production · Ruminants

2.1 Scope and Context

Livestock agriculture contributes to a wide range of environmental and health issues globally. The food system contributes around a third of the total global anthropogenic greenhouse gas emissions (IPCC, 2019a), and livestock production and consumption makes up a large share of these emissions, particularly ruminants (e.g., cattle and sheep). Agriculture occupies about half of all habitable land on Earth, uses about 70% of all global fresh water, and is a main driver of global land use change and land degradation (IPCC, 2019a). Livestock and the production of feed for livestock take up the majority (around 83%) of global agricultural land and lead in some regions to water scarcity issues. Meat and dairy production are further

B. van Oort (✉) · R. Andrew · N. B. Holmelin
CICERO Center for International Climate Research, Oslo, Norway
e-mail: oort@cicero.oslo.no

D. Mayerfeld (ed.), *Our Carbon Hoofprint*, Food and Health,
https://doi.org/10.1007/978-3-031-09023-3_2

pointed out as a main driver of the current 6th mass biodiversity extinction (IPBES, 2019). High consumption of red and processed meat is also associated with an increased risk of total mortality, cardiovascular disease, colorectal cancer, and type-2 diabetes (Richi et al., 2015), as well as depression (Nucci et al., 2020). In return for this, livestock delivers around 18% of the calories and 37% of the protein to the food system for global consumption (Poore & Nemecek, 2018).

Meat is traditionally an important part of the human diet in many regions of the world, contributing important nutrients such as protein, iron, zinc, and vitamins A, B1 and B12 (Richi et al., 2015). Both total and per capita consumption of meat is globally rising and projected to do so for quite some time to come (Godfray et al., 2018). Ruminant livestock also play an important role in producing milk, manure and labor for e.g. ploughing in many farming systems across the world, and is an important source of income for farmers. Meat production can have a cornerstone role in a country's food production and be important for its level of food self-sufficiency. In sensitive regions, meat production may also be more resilient to climate change than plant-based food production, as animals can be shifted to different types of feed in times of crop failure.

Thus, livestock and meat production and its consumption are of significant importance to humans and are often linked to various agricultural, economic, social, land-use, health, climate and biodiversity policies. This leads to conflicting policy goals, tensions, and debate, where many interests and priorities need to be balanced against each other. To enable informed and balanced decisions at different levels in the food system, an in-depth understanding of the many roles of meat and livestock is required.

In this chapter, we will specifically focus on the potential of changes in especially ruminant meat production and consumption in reducing emissions, in meeting national pledges and the targets of the Paris Agreement. This potential can be a combination of limiting emissions from production through changes in management and production methods, and changes in total production. While global greenhouse gas emissions are dominated by fossil fuel emissions (IPCC, 2014), emission reductions of different greenhouse gases in the food system could contribute significantly to the global effort to limit warming. This includes reductions in both the production and consumption of meat. Without that broader view, beyond production emissions, it would be impossible to meet global temperature targets – even if fossil fuel emissions were immediately halted (Clark et al., 2020). To that extent, we will consider ruminant emissions and their effects on global warming in a global perspective.

Naturally, there are regional, national and local differences in livestock related emissions, both in production and in consumption (Gerber, 2013; OECD, 2021; Sans & Combris, 2015). The rates of emissions and the opportunity space to reduce these emissions are therefore context specific. Furthermore, since livestock rearing has broad agricultural, economic, and cultural significance, it cannot be reduced to its climate impact alone. We will therefore discuss meat production, consumption and the opportunities for emission reduction using a case study of Norway as an example of a high-income country that has an agricultural sector largely dominated by livestock production. The reason for selecting a high-income country as a case is based on the generally high levels of meat consumption, and thus the potential for emission reduction through shifts in consumption in such countries. We will start however with a number of overarching issues that are discussed at a global scale, with less attention to regional variations.

Unless otherwise specified, discussions of "meat" production and consumption in this chapter refer to ruminant meat specifically, as this is where most of the livestock climate-related tensions and debates are focused.

2.1.1 Global Emissions and Pledges

Recent IPCC reports (Allen et al., 2018; IPCC, 2019a, 2019b) present results from existing peer reviewed literature on the status and rate of change in global emissions and discuss scenarios and pathways for mitigation. The 1.5 °C report (IPCC, 2018) points out that half a degree difference (between 1.5 °C and 2 °C) poses large challenges to human and natural systems, and that already with today's warming at about 1 °C above preindustrial levels many of these systems are challenged. In other words, the Paris Agreement – which sets out to limit global warming to well below 2 °C and pursue efforts to limit it to 1.5 °C above preindustrial levels – is not an ideal or desired world to aim for but may be the most ambitious target attainable under the circumstances. Meanwhile, emissions of all greenhouse gases are increasing (Fig. 2.1), national pledges fall short of the Paris Agreement targets, and national

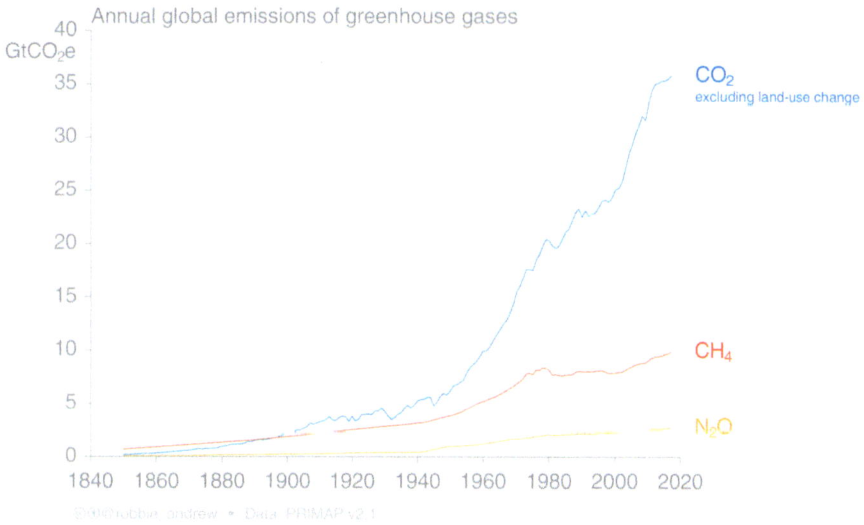

Fig. 2.1 Historic and current global emissions of greenhouse gases CO_2, CH_4 (methane) and N_2O (nitrous oxide), based on IPCC AR5 100-year GWPs. (Source: Gütschow et al., 2016, 2019)

actions and projected policies fall short of the national pledges (Climate Action Tracker, no date). An important consideration however is the ratcheting-up mechanism of the Paris Agreement: while countries may not meet their pledges and pledges may not meet the Paris Agreement now, the inbuilt increased effort may change this situation. However, climate action is urgent, and the longer we wait with emission reductions, the more difficult it will become to meet the targets.

The IPCC (IPCC, 2019a) estimates that around 23% of the total anthropogenic greenhouse gases come from the combined agriculture, forestry and other land use emissions. The most important greenhouse gases from agriculture are nitrous oxide (N_2O) and methane (CH_4). Nitrous oxide increases especially reflect the great increase in the use of inorganic nitrogen fertilizer since 1960, a large share of which is used for cultivation of livestock feed. Drivers of methane emission increases are more complex. Recent studies show increases in both methane emissions and atmospheric concentrations (Nisbet et al., 2019; Saunois et al., 2019), and since 2012 methane emission rates follow the high emission scenarios from the IPCC.

Until recently, the causes of the rise in atmospheric concentration of methane were not fully understood. Both an increase in emissions and a decrease in its atmospheric breakdown have been pointed out as possible causes for the analyzed period, which means that methane would have a longer atmospheric presence and increasingly contribute to warming (Nisbet et al., 2019). Ni & Groffman (2018) find that there are declines in methane uptake in forest soils, which may contribute to the atmospheric increase. While there are multiple sources of methane emissions, agriculture and waste – with enteric fermentation and manure handling as main contributors – are identified as the main anthropogenic sources (Saunois et al., 2019). Indeed, Saunois et al. (2019) find that enteric fermentation and manure handling emissions have been increasing since 2000, which corresponds to an increase in the

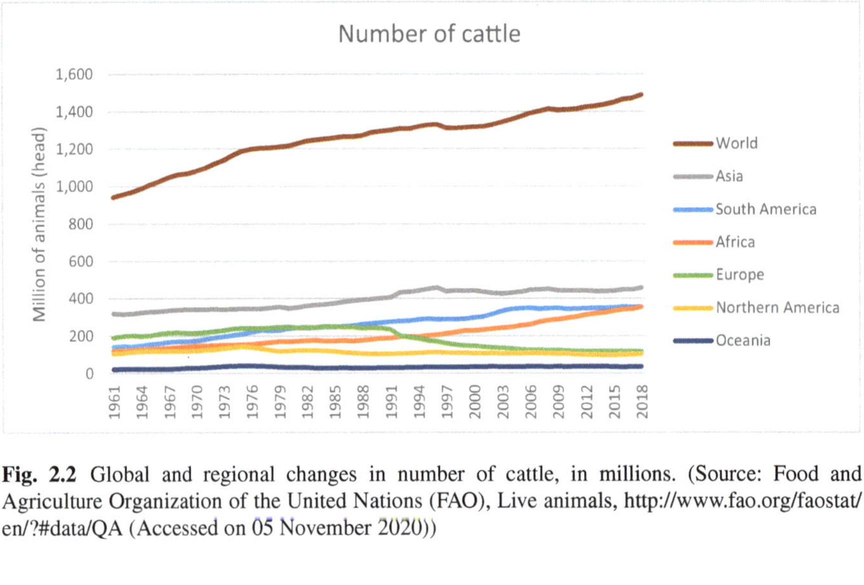

Fig. 2.2 Global and regional changes in number of cattle, in millions. (Source: Food and Agriculture Organization of the United Nations (FAO), Live animals, http://www.fao.org/faostat/en/?#data/QA (Accessed on 05 November 2020))

global number of ruminants (Fig. 2.2), especially in regions with relatively higher emissions from ruminant production (Gerber, 2013; Saunois et al., 2019). A recent paper by Chandra et al. (2021) furthermore shows that reductions in CH_4 growth rates in the 1990's were due to reduced emissions particularly from oil and gas exploitation and enteric fermentation emissions from Europe and Russia, while the resumed growth of CH_4 from 2007 can be attributed to increases in emissions from coal mining mainly in China and intensification of ruminant farming in tropical regions.

2.1.2 Global Meat and Livestock Emissions

The global food system is estimated to contribute between 21% and 37% of global emissions (IPCC, 2019a). More than half of these emissions are connected to livestock and fish production. Direct emissions from livestock and fish farm production, which includes enteric fermentation, manure- and pasture management and fuel use for fish farms, contribute around 31% to the total food system emissions. Indirect emissions contribute an additional 6% from feed crop production, and 16% from land use change emissions, which amounts to around 53% of the total food system emissions (Poore & Nemecek, 2018; Ritchie, 2019). Estimates put the livestock supply chain's share at around 14.5% of total annual global greenhouse gas emissions (Gerber, 2013; IPCC, 2019a).

Livestock contributes substantially to the emission of these gases, with respectively 44%, 53% and 5% of the total global anthropogenic methane, nitrous oxide and CO_2 emissions (IPCC, 2019a). Cattle are the livestock sector's biggest contributors with 65% of the sector's emissions, while pigs, poultry, buffaloes and small ruminants each represent between 7% and 10% of the sector's emissions (Gerber, 2013). The reasons are firstly, that cattle are the most numerous ruminant livestock with around 1.5 billion animals globally, while there are about one billion sheep and about as many goats (FAOSTAT, 2020). Secondly, cattle are larger animals and like all ruminants have digestive characteristics that result in higher emissions from enteric fermentation (Saunois et al., 2019).

The reason methane is much discussed in the context of ruminant emissions is because when using the GWP_{100} metric, methane warms the atmosphere about 28 times more than CO_2, and because it makes up such a large proportion of the total emissions from ruminants. Methane is different from other greenhouse gases such as CO_2 and nitrous oxide in that it stays for a much shorter time in the atmosphere, around a decade on average, compared to more than a hundred years for nitrous oxide and over a thousand years for CO_2. Besides being a greenhouse gas, methane breaks down to CO_2, water vapor and ozone, and near-ground ozone contributes to air pollution which negatively affects human health and crop growth (e.g. Sillmann et al., 2021; Monks et al., 2015). The contribution of the different gases to warming therefore changes over time and mitigating these different gases through actions across sectors similarly has different effects on warming.

Around 39% of the global livestock supply-chain emissions come from enteric fermentation methane (Gerber, 2013), while for cattle alone this share is estimated at around 42.6–46.5% (Opio et al., 2013). Saunois et al. (2019) find that enteric fermentation and manure make up 54% of total agriculture and waste methane emissions, and 30% of the total anthropogenic methane emissions. This corresponds to similar findings in a recent study (Chang et al., 2019) which suggests that the share of enteric methane emissions from livestock to the total atmospheric methane concentration is even higher than previously assumed using a more refined way of calculating the relative contribution of C3 and C4 plants in feed.

Thus, the food system, livestock and ruminants contribute considerably to global emissions. This suggests that there is both an opportunity and a need for reducing emissions from the food system. By considering alternatives for production and consumption, a substantial fraction of these emissions could be reduced, and a sector that currently contributes to warming could help compensate hard-to-mitigate emissions from other sectors.

2.2 Livestock Supply-Chain Emissions and Regional Differences

Around 95% of the total global livestock supply-chain emissions come from enteric emissions, feed production, land-use change, fertilizer use and manure management (Gerber, 2013). Feed production (including pasture management, fertilizer production and use, and the expansion of pasture and feed crops into forests – see Gerber for specifications) and processing, and enteric fermentation from ruminants are the two main sources of emissions, representing 45% and 39% of sector emissions, respectively. Manure storage and processing represent 10%. Post-farm emissions and direct and indirect energy use make up the remaining approximate 5%. The distribution is similar for milk production from cattle, with post-farm emissions making up about 6%, but only about 0.5% of beef production emissions are post-farm (Gerber, 2013).

There is a distinct difference between the emission intensity per kilogram (kg) of product from beef produced from dedicated beef cattle and beef produced from dairy cattle, primarily because the emissions from dairy cattle are divided between milk and meat, while specialized beef herds produce only beef. At the global level, this means that specialized beef has almost four times higher emissions than dairy beef. In addition, pasture-fed beef tends to give higher emissions per kg product than beef from animals that receive a higher proportion of concentrate feed and spend more time indoors, mainly because of feed quality and herd management. The reason for this is that grass has a higher cellulose content making it less digestible than concentrate feed; it also has less carbohydrate. This means that grass-fed animals are less efficient, grow more slowly, and live longer before reaching their desired slaughter weight, resulting in more emissions per kilogram of product. Secondly, methane emissions from the breakdown of grass are higher because the rumen bacteria breaking down cellulose produce higher concentrations of methane than the rumen bacteria breaking down non-cellulose materials. In sum, grass-fed animals produce more methane, in absolute terms and per kg product (Garnett et al., 2017). On a per-animal basis, the story may be the reverse. While roughage – which gives higher emissions – makes up a larger proportion in beef cattle feed (in Norway: 93% for suckling cows, 61% for bulls) than in dairy cattle (in Norway: 55%), dairy cows eat about the same total amount of roughage, and concentrate consumption comes on top of this (Animalia, 2020). This means that the emissions from concentrate production come on top of other emissions for dairy cows, resulting in higher total emissions on a per-animal basis. But, as explained, when distributed over lifetime and production, emissions per kg product end up lower due to the distribution of emissions over milk and meat.

The global emission intensities – the emissions per kg product – and distribution between products vary greatly between producers and regions, with agro-ecological conditions, farming practices, animal productivity, division of emissions over single or multiple products (e.g., milk and meat), land-use change related to feed production, and supply chain management explaining the heterogeneity observed both within and across production systems. European countries generally come out with relatively low emissions, due to high productivity, multiple products to allocate emissions to, and emission-reduction measures. Latin America, South Asia, East- and Southeast Asia, and South Sahel Africa come out with generally higher emissions per kilogram product (e.g. Caro et al., 2014; Gerber, 2013), and are also identified as the main regions with increasing methane emissions from enteric fermentation (Chandra et al., 2021; Chang et al., 2019). Asia, South America and Africa are also the regions with highest – and increasing – number of cattle (Fig. 2.2). Although the emission intensity is decreasing both in developing and developed countries for most meat types including beef, pork, and chicken, Caro et al. (2014) calculate that this does not compensate for the increasing numbers of these animals.

Higher emissions are especially due to higher enteric methane emissions, reflecting low feed digestibility which leads to higher enteric and manure emissions, a lower productivity and slaughter weight, and animals living longer before slaughter, thus emitting for a longer time. In addition, emissions related to land-use change such as pasture expansion into forested areas play a large role in Latin America. In these regions, a great share of the cattle are specialized beef herds, such that emissions are mainly allocated to one product: meat. In Europe, in contrast, about 80% of the beef is produced from dairy animals (both calves and dairy cows), such that emissions are distributed between meat and milk products, and therefore give lower emissions per kg meat.

2.3 The Role of Metrics

The importance of methane is reflected in emission assessments for individual foods: meats from ruminants are generally found to have the largest 'carbon footprints', with enteric methane a major contributor. As a result, methane emissions dominate in the argument over the climate impacts of ruminants. It is important to understand that when comparing for example between foods, exactly how one compares different greenhouse gases and converts methane to so-called 'CO_2 equivalents' has an important role in understanding climate impacts. While this difficulty in comparing between different gases using a metric called "global warming potential" (GWP) was already pointed out in the first IPCC reports, this debate has recently been rekindled with much attention and different opinions on what this means for livestock production and emission reductions.

The background for this is that GWP_{100}, the metric that globally has been adopted as the standard for reporting on national emissions to the United Nations Framework Convention on Climate Change (UNFCCC), takes a hundred-year perspective, and as such misrepresents the impact of short-lived climate gases such as methane on warming over time (Cain, 2018). An alternative metric called GWP* (pronounced 'GWP star') was proposed by Allen et al. (2018). This metric focuses not on the total emission over time, but on the rate of change in emissions, thereby linking emissions to their warming effect. It thereby better accounts for methane's strong effect on warming during its short lifetime, and for its reduced effect on warming after its degradation. Key in this new way of calculating methane's effect on warming is that methane does not accumulate in the atmosphere to the same extent as CO_2 does. Due to its shorter lifetime, a near-stable (reducing by 0.3% per year) emission of methane to the atmosphere does not contribute to increased warming. Finally, the greater impact of *changes* to methane emissions are reflected by the GWP* metric, as declining methane emissions leads to a cooling effect not captured by the standard GWP_{100} method, while increasing methane emissions cause substantial warming. Figure 2.3 explains these dynamics in more detail:

While there is nothing new here for the atmospheric science community, the introduction of GWP* has caused a great deal of discussion within the food sustainability community. Especially the insight that "ongoing but stable methane emissions can be compatible with temperature stabilization has been used by some stakeholders to argue that ruminants, particularly those in grazing systems, are (variously) not a problem, or inherently less problematic than monogastric pigs and poultry", as stated in an excellent summary report by Lynch et al. (2020). This report concisely presents how the different metrics compare, and why using a different metric does not mean that we can ignore ruminant methane emissions, or that keeping ruminant methane emissions stable will somehow solve the problem of globally increasing emissions (Fig. 2.1). Some of the key points of this report are that a narrow focus on methane falsely insinuates that ruminant emissions can be reduced to methane emissions, when in fact ruminants are responsible for other climate-changing emissions as well: nitrous oxide and carbon dioxide – from the animals' manure, feed and input production and from land-use change, including deforestation and the clearance of other vegetation for pasture and feed production.

Lynch et al. (2020) show how, aggregated globally, total non-methane emissions from ruminants are still greater than emissions from monogastric livestock. Even when scaling emissions per kg of protein they show that ruminant production can have similar or greater emission intensities of CO_2 and N_2O. Combining these CO_2

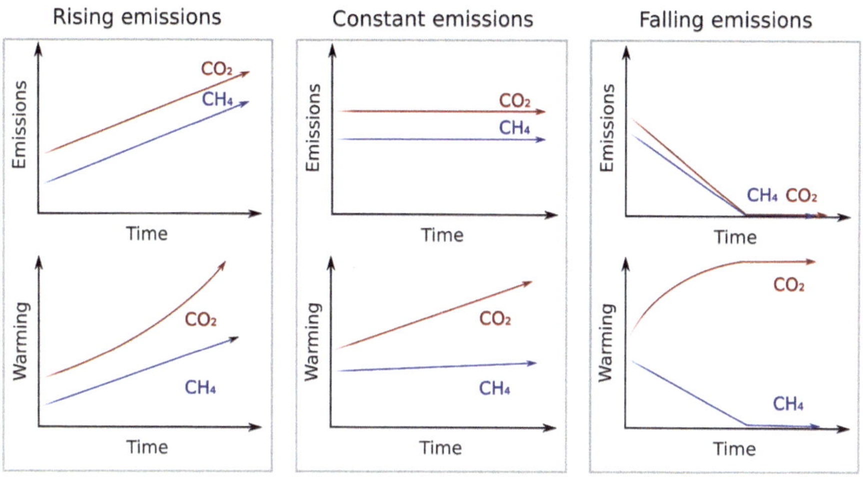

Fig. 2.3 Temperature responses to various emission scenarios for CO_2 and methane. Left panels: When emissions are rising, CO_2 and methane both cause increased warming. Middle panels: Temperature continues to rise under constant emissions of CO_2, as CO_2 continues to accumulate. In contrast, constant methane emissions lead to constant methane concentrations in the atmosphere, which hold temperature at an elevated, but nearly stable, level, rising only very slowly due to continued multi-century adjustment of the climate system. Right panels: Temperature continues to rise in response to falling CO_2 emissions, as long as they remain above zero, but temperature falls in response to rapidly falling methane emissions. When emissions reach zero, the temperature response to CO_2 remains constant for many decades at whatever level it has reached due to cumulative CO_2 emissions over the entire industrial period, while the temperature response to methane declines to near zero within about a decade, because of its short lifetime. (Source: Allen et al., 2018)

and N_2O emissions for the case of the UK using the 100-year global warming potential (which, as illustrated and discussed in Lynch et al. (2020), provides a more direct equivalence for these gases than for methane over important policy timeframes) suggests that ruminants can still have a greater climate impact per kg protein produced than monogastric livestock, even without considering the role of methane. Using and thinking around different metrics and reporting methods is important to reflect the actual warming effects of ruminants and might also find climatically sustainable ways to continue ruminant production under certain conditions. Lynch et al. (2020) conclude that this climatic space is still limited, and not compatible with ever-increasing demand for ruminant products. This also agrees with Ridoutt (2020), who find that when using radiative forcing (RF) as a metric for the footprint of sheep in Australia, this production could be "net zero" and even contribute to climate cooling, but only because the production is in overall decline.

In short, regardless of which metric is chosen to compare methane with CO_2, current consumption and production methods of ruminant livestock continue to be important contributors to the problem of climate change and to other negative environmental impacts, which was also demonstrated directly by use of a climate model – thereby bypassing the question of metrics – by Reisinger and Clark (2018). Even if near-constant methane emissions associated with their stable consumption in a given region does not result in additional warming, stabilization alone does not absolve methane's contribution to warming: the fact that those methane emissions continue means the world is warmer than it would otherwise be (Rogelj & Schleussner, 2019). The level of stabilization has important consequences for how much methane continues to contribute.

2.3.1 Biogenic Emissions

It is often argued that emissions from livestock are biogenic, in contrast to thermogenic or pyrogenic, such as for example methane from fossil fuels. That they are biogenic means that they are part of the fast carbon cycle with a relatively rapid turnover, such as where animals eat plants and plants take up carbon from the atmosphere. Indeed, several natural processes including animal digestion produce such biogenic methane, but each of the three process categories has both anthropogenic and natural components. The global methane budget (Global Carbon Project (GCP), 2021; Saunois et al., 2019) calculates and presents methane emissions across these categories. Due to the relatively fast cycle of biogenic methane, it is sometimes argued that in sum these emissions "don't count" in global warming since they are part of the atmospheric carbon cycle, and not newly added to the atmosphere from deeply buried fossil sources (CLEAR Center at UC Davis, 2020).

While such arguments make an important and valid point about the difference in biogenic and thermogenic methane, and in fact mention the importance to reduce methane from all sources, this point tends to be used to justify beef production. As such, it distracts from the many emissions coupled to beef production, including land use change, fertilizer and pesticide production for feed, heating of buildings, and the fossil fuel use for many of the post-production parts in the value chain related to packaging, transport, etc. While methane has a short half-life of around 9–12 years, beef production also involves emission of long-lived greenhouse gases CO_2 and nitrous oxide which will accumulate in the atmosphere and contribute to warming over a long time. Secondly, while methane is in the atmosphere, it is a much more powerful greenhouse gas than CO_2 and contributes much more to warming over its short lifetime than CO_2 over the same period. Reisinger and Clark (2018) found that non-CO_2 livestock emissions cause around 19% of the warming that the world is experiencing now, and 23% when emissions from conversion of land to pasture were included. Finally, both the number of cattle (Fig. 2.2) and global consumption (Fig. 2.10) of beef are increasing. While production is becoming more efficient, more cattle and more beef means more emissions, of all gases connected to beef production. And as Fig. 2.3 shows: increased CO_2 and CH_4 means increased warming.

2.4 Norway

Direct emissions from livestock are only part of the story. In order to have a holistic understanding of the climate impact of meat production one must also account for other factors influencing overall climate impact, both positively and negatively. These include the potential role of ruminants on soil carbon via grazing activities, and the positive effect on reflection of solar radiation (albedo) of these grazing activities. Climate impact is determined by the sum of emissions, uptake and storage of different gases, and of the role of radiation reflection. Emissions further depend on the demand for livestock products. All of these may differ spatially (different for global regions and/or altitudes) and temporally (time horizon over which the effect takes place). Some of these effects can be reversible, and thus contribute differently to the overall effect over time. Since these effects can differ greatly between locations and over time, we will discuss these specifically for the case of Norway, followed by a more general discussion later in this chapter.

2.4.1 Agricultural Production, Area and Emissions

Agriculture is of great economic and cultural importance in several provinces in Norway, and Norway's land-based food production has a strong livestock focus. At the national level, the meat industry employs more people than the entire Norwegian fishing industry (AgriAnalyse, 2019a), though there is considerably more money in the latter. Despite its local importance, only around 2.5% of Norway's total land area allows for agricultural production, due to climatic conditions, soil properties and much mountainous terrain. Of this small area, around 70% is used for grassland, which serves as feed for ruminants, while around 30% of the area allows for grain, vegetable and fruit production, but mostly grain (Fig. 2.4). On average, around 80% of the annual grain yields are used as livestock feed (Knutsen, 2018), due to annual weather variations which determine the protein contents in the grains, to the protein level criteria set by the mills, and to the small price difference between food grain and feed grains making it relatively risky and unprofitable to invest much in food grain production. All in all, around 90% of Norway's available agricultural land is used for livestock production. In addition to that come the extensive, outfield pastures which are not included in this area.

Consequently, the national agricultural emissions are dominated by livestock production. Agricultural emissions make up around 8.5% (in 2019) of total national emissions, not including emission from wetland drainage for feed production. Most of the agricultural emissions come from methane (58%) and nitrous oxide (40%), primarily linked to livestock production. Livestock emissions – methane from digestive processes (51%), and methane and nitrous oxide emissions from manure (around 20%) – add up to around 71% of the agricultural emissions (Norwegian Environment Agency & Statistics Norway, 2021; Rivedal, 2020). Livestock's share

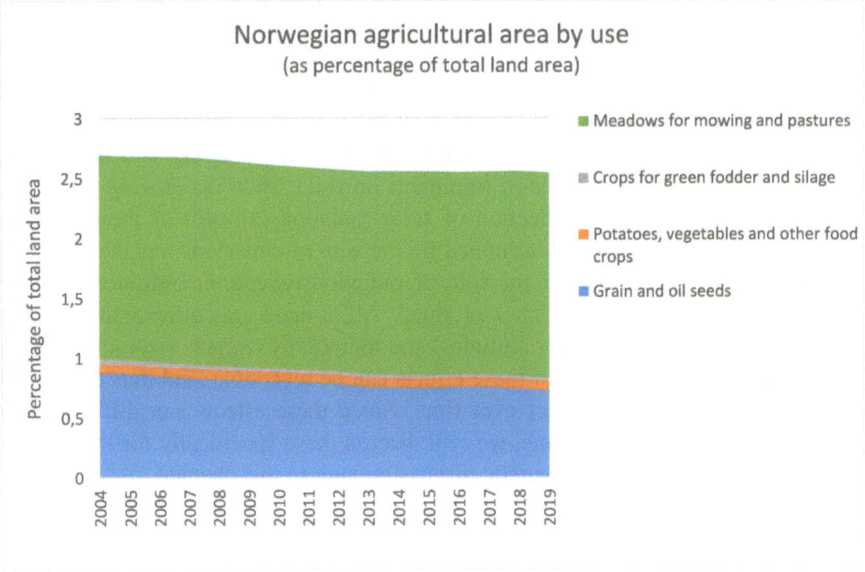

Fig. 2.4 Norwegian agricultural land use as a percentage of total land area. Extensive pastures are not included. (Source: Statistics Norway (ssb.no), table 11506)

would be larger if also including emissions for feed production, related land use and land use change, and energy use. Overall, around 56% of the national methane emissions come from agriculture and livestock, while the rest of Norway's methane emissions come from the oil and gas industry, heating and landfill emissions (Statistics Norway, 2021a). Thus, livestock and specifically ruminants contribute significantly to the national emissions of methane and nitrous oxide.

An overview of the last three decades of Norwegian livestock production (Fig. 2.5) shows that the numbers of both sheep and cattle have declined, while pork and chicken production have seen a strong increase over time. The reduction in cattle is mostly due to a strong increase in milk production per cow from around 6000 to 8000 kg milk per cow per year from 2000 to 2018. This increase in productivity combined with a declining demand for milk has led to a reduction of the total number of dairy cows. This also led to reduced meat production from dairy cows, which – given sustained beef consumption – has in part been compensated for by an increase in suckling (i.e., beef) cows to maintain domestic beef supply (AgriAnalyse, 2019b). The increase in chicken production also corresponds to feedback from producers reporting a much higher profitability in poultry and egg production compared to milk producers and beef producers (AgriAnalyse, 2019a).

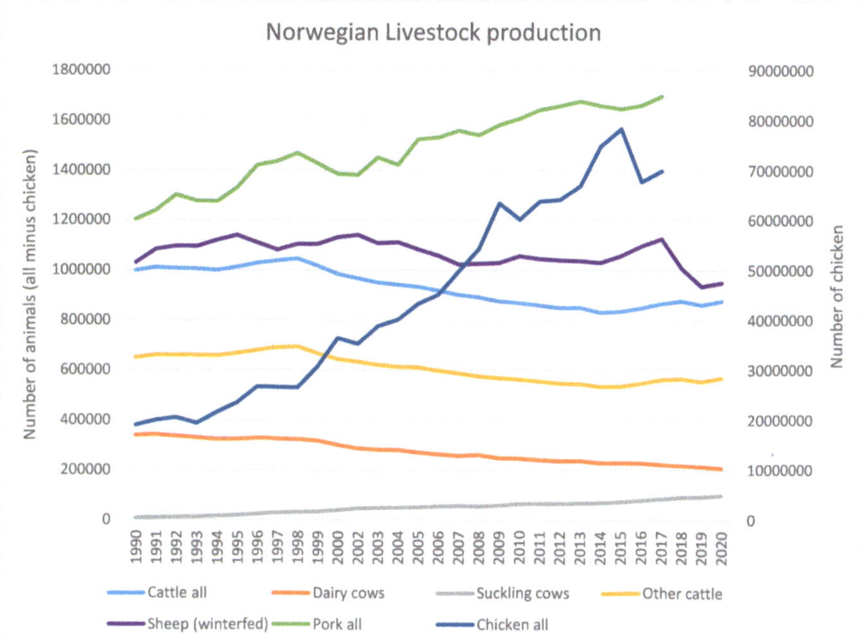

Fig. 2.5 Norwegian livestock production. (Source: Statistics Norway, 2021c. "Other cattle" in this figure consists of calves under 1 year old, males in classes 1–2 year old and over 2 year old, and heifers 1–2 year old and over 2 years old (Statistics Norway, 1999))

2.4.2 Carbon Uptake and Grazing Effects on Albedo

Soil carbon in Norway: For Norway, data on soil carbon saturation and potential for uptake are sparse or not available. Both the landscape and vegetation in grazing areas are heterogenous, making it difficult to give a good estimate for the soil carbon storage potential. Using Swedish data for soil carbon storage as an example, Hillestad (2019) calculates that for Norway, carbon storage in soil due to grazing could hypothetically offset about a third of the emissions from grazing livestock. While this is just a hypothetical, not necessarily reflecting realistic Norwegian conditions, the authors state that the calculation does serve as a reminder that there may be a large carbon storage potential which is currently left out of the emission equation for grazing ruminants.

A recent report on the carbon storage potential in Norwegian soils (Rasse et al., 2019) points out that around 70% of the agricultural land is used for grass production, and this soil is considered to already be rich in organic carbon. Similar to other studies (e.g. Garnett et al., 2017), the authors state that being close to a saturation point, the potential for further carbon uptake is likely limited. Extensive measures would be required to keep levels of carbon in the soil above this equilibrium. Furthermore, as soil temperatures increase with global warming, decomposition of

Fig. 2.6 Norwegian cattle in extensive pasture. (Source: "Storfe på fjellbeite i Frydalen" by Øyvind Holmstad, https://commons.wikimedia.org/wiki/File:Storfe_p%C3%A5_fjellbeite_i_Frydalen.jpg. Changes to original: adjusted lighting)

soil organic material increases, an effect which would be particularly strong in the Norwegian setting with a cold climate and high soil carbon content. A transition from food crop production – where soil carbon levels are lower because of continuous removal of biomass through harvest – to grass production would increase soil carbon uptake (Klimautvalget, 2016). However, current policies aiming at optimizing agricultural land for food production discourage such transitions. Land conversions aimed at increasing grass production largely consist of transforming wetland to grassland. This process however leads to large emissions of CO_2 from carbon-rich wetlands, and moreover is limited through policies setting an annual limit to such conversions. Furthermore, a recently passed law aims at preventing such conversions altogether (Farstad et al., 2020).

Pastures are used for grazing, in contrast to meadows which are primarily being mown to make hay for animal fodder. Pastures can be separated into intensive grasslands or infield pastures, which are being managed through agricultural practices such as seeding and fertilizing, and extensive or outfield pastures or rangelands, which primarily grow native vegetation and are extensively managed through regulated intensity of grazing. Due to structural changes in the farming system linked to low profitability (AgriAnalyse, 2019a), the number of farms has been reduced, while the average area per farm has increased. Between 2000 and 2017, the total number of farms has been reduced from around 70.000 to around 40.000, while farms with cattle and sheep have decreased particularly over that time, by roughly

50% and 30%, respectively. In this process, especially the number of small farms, which traditionally make greater use of extensive pastures, has decreased. As a result, areas of farmland linked to small farms have gone out of production and the role of rangelands or extensive pastures for grazing and their potential role in soil carbon storage has decreased over time. This is a process that has gone on over a long time: The share of extensive pastures in total feed uptake for cattle decreased from around 58% in 1939 to 27% in 1996, while sheep increased their share of extensive pasture feed from 31% to 70% over the same period (Hegrenes & Asheim, 2006). The same report calculates that in 2004 the average duration on extensive pastures for cattle was between 60 and 90 days per year, and their use of extensive pastures has been reduced to around 12%–14% of total feed intake. For sheep and lambs this is calculated at respectively 26% and 42%. Most of the feed for ruminants thus comes from grazing on intensive pastures, cultivated roughage harvested from meadows, and concentrate feed – and the contribution from grazing on extensive pastures is being reduced – as therefore is the potential of soil carbon storage through grazing.

Combined with climate change effects promoting vegetation growth, outfield pastures have seen an increase in shrubification (Rasse et al., 2019). Increased growth of shrubs and bushes leads to increased carbon storage above ground, but it is unclear how this affects long-term carbon storage below ground. Current understanding is that the total carbon storage is higher in areas with shrubs and bushes than in grazed outfield pastures, because such shrubified areas contain both a higher above-ground carbon storage and a legacy of high soil carbon. This is supported by Schmitz et al. (2014) who point out that reduced browsing by wild moose in a boreal forest could lead to greatly increased carbon storage in the form of increased primary production. Speed et al. (2014) similarly found that areas with high sheep-grazing pressure had a different vegetation composition and lower total carbon than areas without grazing sheep. A reduction in grazing resulted in establishment of birch forest and a much higher total carbon storage compared to the grazing situation (Speed et al., 2014), while an intermediate grazing pressure led to increased soil carbon (Martinsen et al., 2011). Regrowth after grazing ceased was found not to change the total carbon but move the carbon balance from soil to above ground vegetation (Speed et al., 2014).

Other studies point out that while maintaining low densities of e.g. sheep, increased grazing in productive grassland systems is expected to lead to reduced soil carbon storage, with soil biota more active due to grazing, leading to a high bioturbation and increased degradation of organic material (e.g. Austrheim et al., 2016). On the other hand, less productive systems dominated by browsers (such as moose and deer) instead of grazers (such as sheep and cattle) will have a negative effect on soil organisms and slower degradation of organic material, which would be positive for soil carbon storage. In addition to the direct and indirect effects of grazing on emissions, it is also important to take emissions from fertilization of intensive pastures into account (Grønlund et al., 2008), as use of extensive pastures only makes up a small overall proportion of the total annual feed intake of cattle. Finally, while ruminants are able to convert grass to food, and grass can be a lower cost resource than concentrate feed, this does not mean that grass-fed meat emissions are lower. As discussed earlier, grass is less digestible than concentrates and it is the grass-degrading micro-organisms in the rumen that produce methane. Secondly, because concentrates contain more sugars than grass, animals fed concentrates grow faster and reach their slaughter weight earlier than animals reared on roughage. Both factors lead to higher methane emissions for grass-fed animals compared to concentrate-fed animals. In Norway, roughage makes up a larger proportion in beef cattle feed (93% for suckling cows, 61% for bulls than in dairy cattle (55%) (Animalia, 2020). Steers consume a higher (85%) percentage roughage, but make up only about 0.7% of the total slaughter volume in Norway (Een Thuen & Tufte, 2019). However, dairy cows eat about the same total amount of roughage as suckling cows, and their concentrate consumption comes in addition to this (Animalia, 2020). This means that the emissions from concentrate production come on top of other emissions for dairy cows, resulting in higher total emissions on a per-animal basis. Around 70% of Norwegian produced beef currently comes from dairy cow production, and 30% comes from suckling cows. For dairy cattle however, emissions are distributed over both milk and meat, and when distributed over lifetime and production, emissions per kg beef end up lower.

While the sum of grazing, emissions and carbon uptake in Norway is not known, as there is no published literature on these combined effects from beef production on extensive pastures in Norway, our current understanding is that regrowing of extensive pastures results in higher total carbon storage, and from a climate point of view increased grazing is not an optimal emission reduction measure. Overall, the role of carbon storage through grazing seems limited in Norway. Cattle spend a limited time each year outdoors, extensive pasture grazing adds only a small proportion to the annual feed, and there are contrasting findings on the role of grazing on carbon storage or release. Roughage leads to reduced production and higher emissions in the digestion process compared to concentrate feed. Depending especially on the land use conversion emissions needed for feed production, such as tropical deforestation for soy production, or wetland conversion for grass production, total emissions related to beef or milk production can increase further. Norway only imports certified deforestation-free soy, and wetland conversion is being limited, suggesting land-use change related emissions are limited. Forest regrowth could moreover contribute to a higher total carbon storage than in grass soils alone. On top of this we need to consider that carbon accumulation in soil is time limited, as the soil would reach a saturation point, and that this process is reversible. Thus, Rasse et al. (2019) conclude that "a less ambitious but more attainable goal would be to try to avoid losing soil carbon as a result from a warming climate."

Albedo: Albedo is a term used to describe the reflectance of the Earth's surface, with high-albedo surfaces reflecting more sunlight, absorbing less as heat. The highest albedo surfaces are fresh snow and ice, while among the lowest is open ocean (which is why the melting of Arctic sea ice is particularly problematic). Pasture reflects more sunlight than coniferous forest during summer, and, because pasture bears much more snow in winter than do tree canopies and trees shade underlying snow, the albedo of pastures in winter is considerably higher than that of forests (Davidson & Wang, 2005).

Several studies have shown that herbivores can play an important role in regulating the density of vegetation (e.g. Olofsson et al., 2009). Grazing can keep pasture open, and thus contribute to higher albedo. Studies in Norway on the role of herbivores on vegetation and solar radiation effects have focused especially on the effects of grazing by reindeer and small mammals such as voles and lemmings (Olofsson et al., 2013; te Beest et al., 2016). These studies find that reindeer have the largest effect on shrubs in tundra, while voles and lemmings have a larger effect in the forest. Te Beest et al. (2016) find that in Arctic tundra, such albedo changes are first dependent on season, with the largest change taking place in the snow-free summer. Secondly, they find that especially transitions from grasses to shrub tundra drives the difference in summer albedo, rather than transitions from dwarf-shrubs (such as birch trees) to tall-shrub tundra. Thirdly, while such grazing effects can have large effects on albedo, radiation and thus warming, these effects are limited to areas with high reindeer densities, since a dramatic vegetation change is necessary. Also in mountainous vegetation changes in Switzerland it was found that the largest effect of vegetation changes on albedo are connected with relatively large vegetation transitions from open land to open forest but not in later stages of forest development (e.g. Schwaab et al., 2015).

Such considerations can also play a role when discussing the climate effects of ruminant meat production. Climate change and changes in land management lead to a northward-migrating treeline and longer growing seasons in boreal regions (IPCC, 2019a). Increased shrub growth in extensive pastures in the absence of grazing by cattle or sheep can contribute to a reduced albedo or reflection (Speed et al., 2014). Reduced grazing thus far is a consequence mainly of structural changes connected to increased efficiency and low profitability in agriculture, and other considerations that make mostly small farmers change their farming practices and reduce grazing in the extensively used or unmanaged pastures. However, a climate policy promoting reduced livestock and reduced grazing could add to the process of regrowing of pastures, and thus potentially have an unwanted effect on warming due to albedo changes, but also – as noted above – increased carbon storage in above ground biomass.

Exactly how much grazing – especially by cattle – contributes to keeping these areas open is not known. However, the numbers of extensively grazing cattle are far lower than those of sheep, and only about 27% of all cattle in Norway graze in extensive pastures (see Fig. 2.7). Here it must be pointed out that extensive pastures can be a mixture of many types of vegetation and consist both of forested areas and of more open landscape types, such as high-mountain meadows. For sheep, most spend time on the pastures (the figure includes lambs in the number of grazing sheep, while lambs are not counted in the number of animals in March each year). There is also a difference in grazing period, which for cattle is a legal minimum of 8 weeks in summer, and for sheep around 16 weeks. How much longer they graze above this minimum number of weeks depends among other things on the growth and nutrition of vegetation, altitude and latitude of the pasture. While the use of

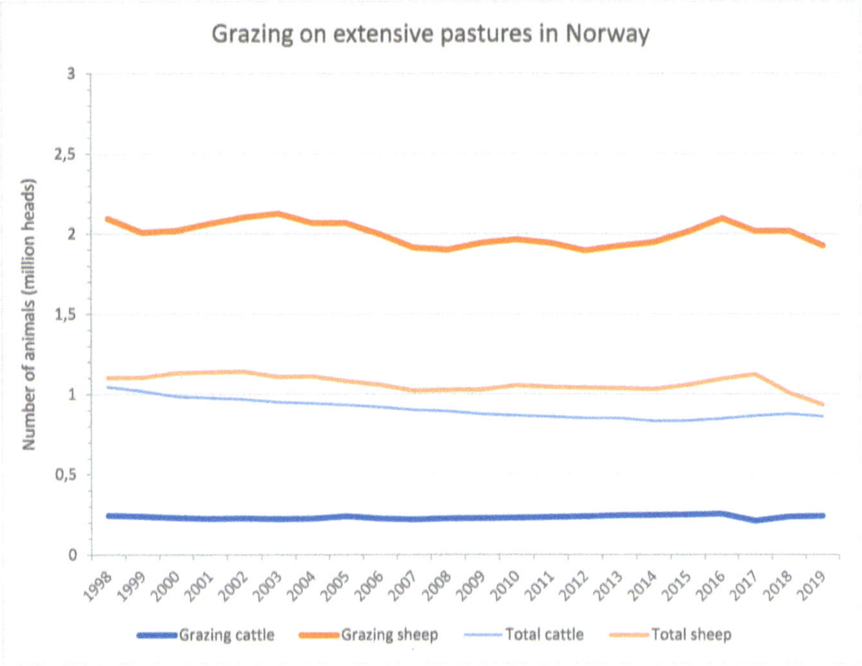

Fig. 2.7 Number of cattle and sheep on extensive pasture area in summer in Norway, and the total number of cattle and sheep (per March 1st) each year. The higher number of grazing sheep than total number of sheep is explained by the counting of spring-born lambs in summer grazing season. (Source: Statistics Norway, 2021b)

extensive pastures is especially relevant for sheep and lambs, individual cattle eat much more than sheep. This increases their relative grazing impact which, based on the likely composition of grazing suckling cows and young bulls and approximate amount of feed each species and age eats (LSB, 2020), is estimated at roughly 70% of the total grazing pressure of sheep and lambs. This is not a negligible amount, but this calculation and the numbers presented in Fig. 2.7 show that (1) not all potential soil carbon storage due to grazing can be attributed to cattle, and (2) if there is carbon storage, this can only be attributed to a 27% fraction of the total number of cattle, most of which are calves/young bulls and suckling cows. These animals as we have seen have higher emissions per kg meat both due to a different feed composition and because they only produce meat, not milk. In this context, about 70% of the beef consumed in Norway currently stems from dairy cattle and their calves, and around 30% from suckling cows and their calves.

While detailed studies of the grazing effect on albedo and warming are thus far lacking, it becomes clear that several factors play a role. For grazing to have a substantial albedo effect there must be a large change in vegetation density *in the absence of grazing*, such as from open landscape to dense forest. Such a substantial vegetation change would require a substantial grazing density. This large vegetation change would further mean a change in albedo, but also in total carbon storage, where a change from pasture to dense forest would mean an increase in total carbon. Some studies (de Wit et al., 2015) suggest that the increased warming effect of a lower albedo of typical boreal birch forest compared to open mountain vegetation, such as outfield pastures, can compensate for the increased carbon storage in forest biomass. This effect depends on the altitude of the change (R. Bright, 2019; R. M. Bright et al., 2016; Schwaab et al., 2015), and on the area needed for grazing and harvesting grass for feeding a cow. However, our understanding based on unpublished data is that - irrespective of altitude - the combined grazing and harvesting for one cow is not enough to prevent regrowth of pasture at sufficient levels that the difference in albedo could offset the warming effect of that cow's methane emissions.

Thus, vegetation changes due to changes in grazing may have both positive and negative effects on albedo and soil carbon storage, and the effects are dependent on several variables, including altitude, type of vegetation, carbon content in the soil, and the role of the animal in question on this complex of variables. Grazing must be intensive and lead to large vegetation changes to have a substantial effect. In Norway, vegetation effects potentially could be large, but there are no data available yet to support that warming effects due to albedo changes may be larger than the combined effects of reduced cattle emissions and increased carbon storage in denser forest.

Overall, open landscape and other partly modified vegetation which is not covered by forest contain relatively high soil carbon concentrations which can be changed by changes in grazing. Both the amount and the stability of carbon can change with reduced grazing and increased vegetation cover. The effects could go either way: a decrease in albedo could contribute to warming if the stored carbon is lower than in an open landscape, but increased carbon storage may also be sufficient to compensate for the change in albedo. Since there are too many unknowns and very local variations in the ultimate effect, Rasse et al. (2019) suggest that the overall focus should be on increasing carbon storage where possible. This potential is largest in areas which currently have a deep soil layer with low carbon content, and lowest in areas with a thin soil layer where an increase is restricted by mountainous and rocky ground. Rasse et al. (2019) further suggest that the key focus therefore must be on avoiding reductions in carbon due to unsustainable grazing practices.

2.4.3 *Food Consumption and Related Emissions*

Norwegian dietary developments are evaluated annually. According to the latest iteration (Helsedirektoratet, 2020), the population still eats less fish, vegetables, fruit, berries, whole grains and fiber than recommended, and too much saturated fat, sugar and salt. Dairy consumption has much reduced over time, while meat consumption has decreased slightly in the last few years (Fig. 2.8) but shows an overall increase over time with different trends for different meat types. A recent report on

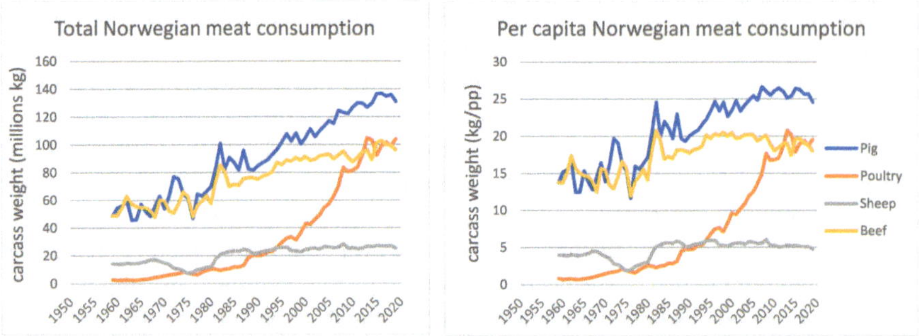

Fig. 2.8 Changes in Norwegian meat consumption over time, in gross kg (left) and gross kg per person (right) per year. Data in gross amounts do not represent the raw weight as bought in the store or the actual intake, but represent carcass weight of meat per year. (Source: Helsedirektoratet, 2020)

Norwegian meat consumption (Animalia, 2020) puts beef consumption in terms of retail weight around 54,000 metric tons in 1990, and shows an increase to around 71,000 metric tons in 2019. While this is a substantial increase, Fig. 2.5 shows that the number of cattle has in fact decreased over time. A gradual increase in the proportion of meat from suckling cows (meat production) has compensated for the decrease in meat from dairy cows. Thus, in line with findings from Caro et al. (2014) for developed countries, more meat from less animals means that production efficiency has increased, and lowered the per kg footprint, also compensating for the total increased consumption of meat.

As an average for the general population, dietary requirements specify a daily energy intake of 2390 kcal, and a daily protein intake of between 60 and 120 grams per day. The national dietary recommendations - which are based solely on health, though processes are underway to include larger sustainability concerns[1] - set a maximum of real intake of 500 grams cooked red and processed meat per week, which corresponds to about 700–750 grams of raw product per week. Red meat here includes both pork and ruminant meat, and this intake is a recommended maximum, not a recommended level of intake. The current average, real intake of red meat in Norway is around 750 gram per week, of which beef and lamb make up about 50%, and pork the other half. These consumption data reflect the average for the population. This means that a proportion of the population eats more than the recommended amounts, while others eat less than this maximum recommended amount. Indeed, the national nutrition survey Norkost 3 (Totland et al., 2012) which presents consumption data from men and women in Norway, shows that women eat on average 620 grams red and processed meat per week, and men on average 1022 grams per week, both raw weight. About 33% of women and 55% of men exceeded the dietary recommendations on red and processed meat, and about 25% of men consumed twice or more than the maximum recommended intake. Thus, for most of the

[1] https://www.norden.org/en/news/nordic-nutrition-recommendations-2022-join-work

population to fall within the recommended maximum intake, the average should be well below this maximum.

While the Norwegian diet is less meat-heavy than that in some other developed countries (Fig. 2.10), its emissions are dominated by animal products, with particularly large contributions from beef and cheese, followed by sheep, pork, milk and poultry (Fig. 2.9). The figure combines emission intensity with volume to show the distribution of greenhouse gases per product for an average person's annual consumption. This results in cases where low emission products contribute high total emissions simply because the consumed volume is high. An example is pork, which has much lower emissions than lamb per kg product but is consumed in much greater quantity. As such, pork contributes about the same total emissions as lamb in the annual diet. Milk likewise ranks highly mostly because of the volume consumed in the country. For beef the reverse is true: per capita consumption of beef is lower than that of pork or chicken (Fig. 2.8), but total emissions are much higher. Also, while Norwegian beef production is more efficient than some countries from which Norway imports beef, such as Africa and Latin-America, consumption from

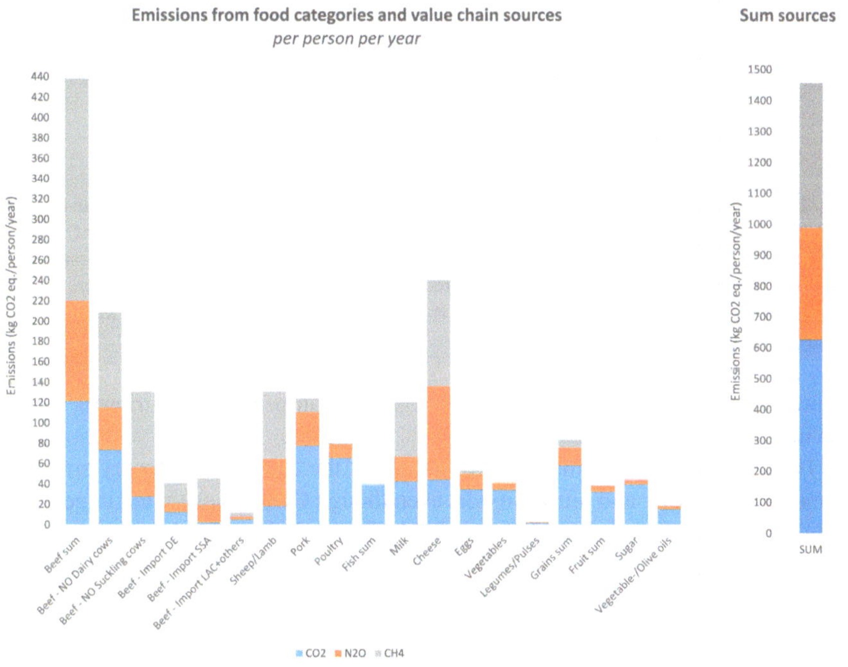

Fig. 2.9 Total emissions from Norwegian annual food consumption per person, for 2017. Consumption is calculated as real intake (gross intake minus food waste) for each food item, including emissions from all value chain sources. Food waste includes all edible food items removed from the value chain, and includes food waste methane, CO_2 and N_2O. Abbreviations in the figure: DE: Germany; SSA: Sub-Saharan Africa; LAC: Latin America and Caribbean. (Source: own calculations)

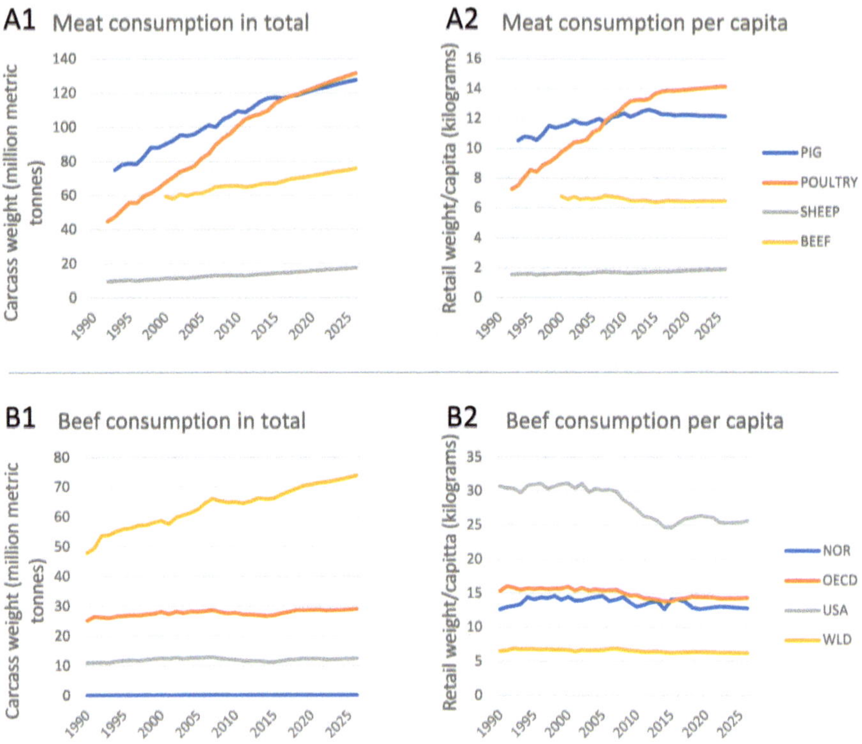

Fig. 2.10 Historic and projected meat (A1 and A2) and beef (B1 and B2) consumption in total (A1 and B1, KT = thousand tons carcass weight) and per capita (A2 and B2, kg retail weight) globally (A1 and A2) and for selected groups of countries (B1 and B2) in the world. Groups presented are the high-income countries represented by the Organisation for Economic Co-operation and Development (OECD); The USA, which also is a part of the OECD; the world as a whole (WLD), and Norway (NOR). (Source: OECD, 2021)

imported beef is minor in most years, and Norwegian beef thus contributes most to the total beef emissions.

The figure also shows that more than 50% of the total annual dietary emissions come from methane and nitrous oxide, even though CO_2 makes up the largest individual share. Methane comes almost solely from ruminants, while nitrous oxide also mostly comes from livestock production, including livestock feed. The figure thus shows that there is considerable opportunity for emission reductions through changes in meat consumption. Changes in consumption moreover would target emissions not only of domestic production but also of imported foods. Of course, substituting high-emission products by low-emission products would increase the volume of the latter, but total emissions would still be significantly reduced.

When discussing changes in consumption and production it is important to be aware of climatic, soil and terrain conditions for production in a country, and secondly how the production system is incentivized through for example subsidies, taxes, food prices or through other policies that touch food production directly or indirectly. We have already discussed the limited agricultural land and opportunities for food production in Norway. However, current agricultural subsidies have a large effect on production as well, with at least 66% of the subsidies going directly to meat or feed production, and around 1.5%–3% going to vegetable and grain production. Also, the small price difference between feed grains and food grains provides farmers with little incentive but higher risks, limiting a switch to food-grain production. There is a general belief that most of the agricultural area can only support grass production, and therefore requires ruminant meat production. Thus, the incentive system tends to stimulate meat production, and is increasingly looking into increasing domestic food and feed production to increase the degree of self-sufficiency.

This focus on increased self-sufficiency has been further fueled by the COVID-19 pandemic, and in the ruminant industry focuses on increased domestic feed production. This means reduced soy imports, with positive climate consequences, but the flip side is that feed needs to be produced domestically, and the question is what kinds of feed should substitute the imported soy. If grass, then this lower energy feed would lead to less productive animals and thus higher emissions especially for milk but also for ruminant meat (Hjelt et al., 2019). Grass moreover is less digestible than concentrates, and it is the grass-degrading micro-organisms in the rumen that produce methane. Thus, unless production drops, methane emissions would increase. Alternatively, if more feed grains or other protein-rich plant products are produced to replace imported soy, this would compete for the little land available for food production and give lower overall food production. In their current strategy for sustainable food production, the Norwegian farmers union proposes to grow plant proteins for feed in rotation with grain production, thus improving soil quality and grain yields and avoiding reduced food production (Norges Bondelag, 2021). Other alternatives, such as extracting proteins from grass in fermenting set-ups, using insect proteins for feed, or using methane-reducing additives could also play a role, but are so far either not tested enough, do not give a large effect, or are difficult to scale up (Rogelj et al., 2018).

In Norway, meat production and market demand have been challenged over recent years, with large overproduction of especially sheep meat. As earlier discussed, especially sheep and lambs, but relatively few cattle, graze in outfield pastures and therefore contribute to some level of carbon storage in soil of unknown magnitude, which is a time-limited and reversible process. There are thus few climate benefits or even market needs of increasing or even maintaining ruminant meat production (although of course there may be other reasons, such as local biodiversity, traditions, local economy, etc.). Exporting any surplus production is out of the question, due to high production costs in Norway, and the fact that export products will not be subsidized, which gives tough competition on the international market.

2.5 Comparison with Other Regions

From the foregoing discussion it becomes clear that emissions are regionally variable, while the climate-change problem with increasing concentrations for different greenhouse gases is a global issue. The role that different nations, regions, the food system, and meat production and consumption play can vary, while all contributions from each of these could help reduce emissions and atmospheric concentrations of greenhouse gases, and the consequences of climate change for all regions and sectors, including food production. In this section we look at some of the Norwegian trends and findings in a broader context to analyze how generalizable they are.

2.5.1 Soil Carbon in a Broader Perspective

Soil carbon has been discussed in the Norwegian context, and the key finding was that the opportunity for carbon storage through grazing appears to be limited in Norway. While data are lacking, there is reason to assume that Norwegian outfield pastures are close to their saturation point. Carbon building in soil thus is time limited, and moreover is a reversible process. Alternative vegetation through regrowing of pastures could contribute to a higher total carbon storage than in grass soils alone. Avoiding the loss of soil carbon as a result of a warming climate seems a more appropriate strategy in the Norwegian context.

Other studies confirm that grazing can contribute to carbon storage, but it is important to understand the counterfactuals – what are the alternatives for grass use, carbon storage with alternative harvesting, carbon storage with other use of the land, etc.. A study on European grasslands found that grazing resulted in a higher net storage of soil carbon compared to machine harvesting (van den Pol et al., 2018). But a recent UK report (FFCC, 2021) with a strong focus on how agroecological farming practices can help reduce emissions, found that grazing cattle and their manure can indeed contribute to reducing climate emissions and add carbon to the soil, but even in the best cases the report found that this carbon storage offsets only 20%–60% of the total emissions from grazing cattle.

In a broader context, claims exist (Savory Institute, 2013) that Holistic management, a form of planned grazing of cattle and sheep, if implemented across the world's grasslands could remove carbon from the atmosphere and store this in the soil. It would be able to do so in quantities that offset the substantial annual emissions from livestock production, all man-made emissions of all greenhouse gases every year, and compensate for all emissions since pre-industrial times and return the atmosphere back to natural levels of CO_2 down from over 400 ppm today to 280 ppm in under 40 years (Savory Institute, 2013). However, the authors also immediately comment that "How realistic these numbers are is unknown, since the bases for them have yet to be tested." Indeed, the claim seems rather exaggerated and has been addressed by several authors: When applied to 5 billion hectares of the world's grasslands (one-third of the Earth's land area), holistic management would absorb 45 billion metric tons of carbon dioxide every year from the atmosphere. This claim was recently addressed comprehensively (Garnett et al., 2017; Nordborg & Röös, 2016). These studies demonstrated an upper limit of 800 million metric tons of soil carbon uptake, far below that of the Savory Institute. Moreover, Garnett et al. (2017) point out that soil carbon uptake is time-limited: absorption rates could not go on forever as soils reach saturation levels.

One reason the numbers differ so much is that the Savory Institute calculates that soil absorbing 45 billion metric tons of CO_2 from the atmosphere would reduce the concentration in the atmosphere by 6 ppm. They further set our current annual carbon emissions at 2.5 ppm/year, presumably derived from how much atmospheric concentrations are increasing each year. This latter number is a mismatch with our current emissions of CO_2 (setting aside other greenhouse gases) of about 40 billion metric tons per year, which is in fact equivalent to a little over 5 ppm/year. The reason for the difference is that about half of our emissions every year are absorbed by the oceans and land systems already. According to a study by Cao and Caldeira (2010), if we somehow removed enough CO_2 from the atmosphere to take it back to pre-industrial levels, then the oceans and land will release vast quantities of stored CO_2. Even if ignoring the role of oceans and land systems in the climate system, and allowing hypothetically that 45 billion metric tons could be removed every year forever, then removing 6 ppm/year from an atmosphere currently at over 410 ppm while combatting continued emissions of just over 5 ppm/year from transport, energy systems, etc. would take almost 200 years before pre-industrial CO_2 levels would be reached.

2.5.1.1 Consumption Patterns

Consumption of meat varies across regions and nations, and even between socio-economic groups and genders. Globally, both production and consumption of meat has increased tremendously over recent decades. Ritchie and Roser (2017) present UN Food and Agriculture Organization (FAO) data in visualizations of the amounts and impacts of different types of meat production and consumption across the world. These data show that globally, meat production has increased by a factor of five from 1961 to 2018, with beef and sheep production doubling over this time. Many studies (e.g. Ritchie & Roser, 2017; Sans & Combris, 2015) show that there is a strong link between income and meat consumption, which can be of concern in terms of increasing global development.

Sans and Combris (2015) present data for world per capita meat consumption patterns from 1961 to 2011. These data show how total global per capita meat consumption on average has almost doubled over this period, and for some countries even increased up to eightfold (China). Much of this increase in demand has been met by pork and poultry production, with ruminant meat consumption staying mostly stable over these decades, but per capita beef consumption especially in Brazil and China has increased. The study discussed that many high-income countries have an animal-based protein consumption that exceed needs, and a reduction in meat consumption was only seen in countries with a high per capita income level that few countries in the world have reached, which can have large consequences for the environment and farmland use as the global population increases and becomes wealthier (e.g. Cole & McCoskey, 2013).

These trends in consumption are also visualized in Fig. 2.10 presenting data from OECD statistics, and are similar to those shown in Fig. 2.8 for Norway. The underlying trends per country can vary strongly. The OECD, USA and Norway show a stabilization of meat consumption since 1990 (Fig. 2.10 B1), while globally meat consumption is increasing (Fig. 2.10 A1). The greatest increase can be attributed to pork, poultry, and beef, with especially developing countries increasing their meat intake (OECD, 2021). Beef was the second largest meat category consumed before the 1990s, but has been overtaken by poultry since. Further details in the OECD online database (OECD, 2021) show that beef consumption is in a stable second place in the United States. As in the Norwegian case, the trends and projections suggest that meat- and beef consumption are flattening out per capita (Fig. 2.10 A2, B2), while total consumption is increasing (Fig. 2.10 A1). The difference in trends between total consumption and per capita reflect both an increase in population and an increase in welfare and demand for meat (Clark et al., 2018). Note that retail weight and carcass weight values are higher than the prepared weight of meat intake, which needs to be used when comparing intake to the dietary recommendations and requirements which vary per age, gender and activity level.

Norway's per capita meat consumption is lower than some high-income countries, but higher than the EU average and many developing countries. Norwegian per capita ruminant meat consumption is comparable to the OECD average, but the OECD and Norway levels are far below some countries such as Argentina, United States, Brazil and Canada. This also means that these countries get a far higher proportion of their energy and proteins from animal sources and consume far above the recommended amount of saturated fat. While there is no recommended maximum level for protein intake, high levels have been linked to health problems such as kidney stones,[2] and reductions are recommended. Thus, while per capita consumption may differ, the trends in Norway are comparable to high-income countries, even though Norway consumes more beef and sheep meat, but less pork and chicken per capita than the EU average.

[2] https://www.health.harvard.edu/nutrition/when-it-comes-to-protein-how-much-is-too-much

2.6 Discussion

2.6.1 *The Role of Meat Production and Consumption in Climate Change Mitigation*

The current level and increasing numbers of livestock continue to contribute significantly to global emissions and climate change. Ruminants and methane play an important role in this. While ruminant production is growing in tropical latitudes, consumption of beef is already high in high-income countries. While beef consumption in high-income countries is declining or stable, the ongoing and globally increasing levels of consumption of ruminant meats contribute to increasing warming by methane and other climate gases (Lynch et al., 2020).

Reductions in emissions of long-lived CO_2 and nitrous oxide are critical, as ongoing emissions accumulate with increasing warming effects. Since methane is short lived, methane mitigation measures can have a very powerful and immediate cooling effect, depending on how they are implemented. Early reductions could contribute to temporary cooling, help keeping pathways from overshooting 1.5 °C on their way to that target and keep climate change from reaching environmental tipping points. Late reductions could serve as an "emergency brake". Both pathways could lower global temperatures by the same amount, but late application must consider the potential risks and damages of temperature overshoot, and the delays and difficulties in sudden methane mitigation measures linked to behavioral change such as reduced meat consumption. Methane emissions are also associated with high costs for society. Shindell et al. (2017) find large economic and health benefits of reductions in methane. The study confirms that the agricultural sector can play an important role through changes in livestock management practices, promoting healthy diets including reduced beef and dairy consumption, and reductions in food waste. They also find that recent trends in methane emissions may have offset much of the societal benefits from a slowdown in the growth rate of carbon dioxide emissions. Overall, their results suggest that methane emission reductions via policies including behavioral changes provide benefits at little or negative net financial cost. This matches well with analyses in Norway on the cost and effect of various climate policy measures, where a reduction in consumption of red and processed meat is among the largest and cheapest measures for the agricultural sector (Miljødirektoratet, 2020). It is worth mentioning again here that ruminant meat production produces more than methane alone, and even without methane, CO_2 and nitrous oxide emissions well exceed those of most other meats or plant-based protein products.

The IPCC analyzed different pathways towards 1.5 °C (IPCC, 2014), and each of these scenarios requires reductions of emission of all gases, although not all gases need to reach zero emissions. Mitigation measures affect several climate gases, with different effects. Alternative pathways towards 1.5 °C differ in their degree of how much they remain below or exceed 1.5 °C on their way to this target. These pathways reflect differences in deployment of mitigation measures and have very different implications for the achievement of sustainable development goals. However, the models that are used for these pathways typically do not include potential emission reduction options in the agricultural sector such as substitution of livestock products with plant-based proteins and cultured meat, use of algae protein in animal feed, inhibitors of methanogen and nitrification, or carbon sequestration in soils. In Norway, the use of earlier harvested roughage has also been calculated to reduce emission from enteric methane (Norges Bondelag, 2020). All of these options could reduce agricultural and land-use emissions in the future, when techniques have been developed, tested and implemented on larger scales (Rogelj et al., 2018).

In a recent paper, Clark et al. (2020) analyzed the need of reducing food-system emissions to stay within the 1.5 °C or 2 °C targets if globally all non-food emissions were to reach zero by 2050 or 2075. In these calculations, they analyze methane emissions as calculated by GWP*. They show that even if fossil fuel emissions were immediately halted, current trends in global food systems would prevent the achievement of the 1.5 °C target and, by the end of the century, threaten the achievement of the 2 °C target. They find that if all non-food emissions are rapidly reduced to zero by 2050, only a full implementation of several combined measures would result in a larger than 67% chance of staying below 1.5 °C. These measures include (1) shifting to a plant-rich diet, (2) reduced calorie intake, (3) increase yields and (4) higher efficiency in production and (5) a 50% reduction in food waste. All other combinations and reduced implementations would place us between 1.5 °C and 2 °C. If a full emission reduction to zero of non-food emissions however were only feasible by 2075, the 2 °C reduction target could only be met by at least a 50% implementation of all five strategies, and not by any single strategy. This shows that meeting the 1.5 °C target requires rapid and ambitious changes to food systems as well as to all non-food sectors. Enteric fermentation during the production of ruminants, manure emissions and land clearing and deforestation related to grazing and feed production are among the major sources of food-system emissions contributing to these projections.

Several studies find that lower consumption of livestock products by 2050 could substantially reduce deforestation and cumulative carbon losses (Clark et al., 2020; Stevanovic et al., 2017; Weindl et al., 2017). The highest abatement of carbon emissions (63–78%) can be achieved if reduced consumption of animal-based products is combined with sustained investments into productivity increases in both plant and animal production (Clark et al., 2020; Weindl et al., 2017). Weindl et al. (2017) finds that intermediate rather than ambitious productivity increases in livestock productivity offer the largest benefits within this measure, as it would halt pasture expansion into forested areas, and reduce land requirement for feed production without negative trade-offs with carbon sequestration in agricultural soils. This and other studies emphasize the importance to integrate demand- and supply-side oriented mitigation strategies to avoid "carbon leakage" where a reduction in meat production combined with persistent demand would lead to import of meat from elsewhere, simply displacing emissions. Studies also emphasize the need for a broader sustainability view that also considers animal well-being, livelihoods, water resources, biodiversity and other considerations (Weindl et al., 2017). As part of that broader view, identification of local production opportunities, linkages with socio-economic, ecosystem and other considerations are needed, and alternatives that contribute to multiple goals simultaneously are needed. In many cases, changes in investments, risk distribution, and innovations based on plant-based production may offer viable new routes without impeding local economy and national food security. Reduced consumption of livestock products by high consumers, combined with development and adoption of measures to reduce emissions from livestock production, could allow for global livestock production at potentially sustainable levels (Lynch et al., 2020; Rogelj et al., 2018).

2.6.2 *Current Policy and National Debate: Opportunities and Barriers for Change*

As part of the European Green Deal, the Farm to Fork (F2F) Strategy (European Commission, 2020) sets out a vision to change the way Europeans value food sustainability and to position Europe as the first climate-neutral continent by 2050. The F2F strategy addresses four key areas of supply-chain management in the food industry: production quantity, production practices, consumption, and waste. Within this framework, it presents ideas on how we will need to work together to achieve better management of the four key topics, such as how future diets will need to change. The final version of the strategy softened its stance on meat compared to the draft versions, as previous proposals to "stop stimulating production or consumption of meat" did not make the final cut of the strategy, and a reference to encouraging people to consume "less meat," was refined to "less red and processed meat". However, the final version maintained staunch support for alternative proteins. Moreover, the strategy still laid out a commitment to "help reduce the environmental and climate impact of animal production". It seems therefore that the current policy focus regarding meat is mostly on reducing emissions by improved production techniques, and soft policy stimulation of changes in consumption.

The situation in Norway is similar, with a similar weakening of the stance on changes in consumption. The former white paper on climate (KLD, 2017) stated that emission reductions from Norwegian food production must be prioritized and include a shift towards increased food production while reducing production of high-emission foods. However, neither the current agricultural sector climate plan (Norges Bondelag, 2020) nor the latest national climate white paper towards 2030 (KLD, 2021) include reductions in ruminant meat production. Still, the latter does point to incentives towards changes in consumption using both harder policy measures such as regulating public procurement and softer measures such as climate information about food choices to consumers. The recent climate white paper (KLD, 2021) states that emission reductions in agriculture should not affect national food security and food production, and that food should be produced based on the available resources. In this argument, food production is set equal to feed and meat production, available resources are set equal to grass and limited agricultural land, and food security assumes unchanged demand for meat. Thus, Norwegian official policy considers that continued meat production is important given its grass resources, and meat production must continue but in concert with changes in demand to avoid increased import and carbon leakage effects – thus also assuming that emissions in Norwegian production are lower than elsewhere or can be made so.

The white papers on climate and on agriculture (KLD, 2021; Norges Bondelag, 2020) underline the dilemma of meeting the emission-reduction targets while at the same time meeting the four main agri-political targets of increasing food security, securing agricultural land use across the whole country, increasing value creation and securing sustainable, low-emission agriculture. With such a wide range of food-related policy targets, which are regulated by different ministries, the various policies can conflict. Such policy conflicts are here presented for the case of Norway, but the issue is global and comparable for many countries. The goals and related policies fall under different ministries. As a result, policies are the consequence of priority setting, ministerial power, interdepartmental agreements and trade-offs (Amundsen & Hermansen, 2020; Skagen, 2020). This complex system explains why decisions around reducing emissions from livestock (or otherwise) are no easy task, indeed requiring a holistic approach considering multiple aspects of livestock production and are by no means regulated by climate considerations alone.

2.6.2.1 Policy Versus Public Debate

The national public debate around the climate effects of meat consumption has on the one hand moved between the role of meat production and consumption in greenhouse gas emissions and climate change, and the environmental, biodiversity and health concerns linked to ruminant farming and meat consumption. On the other hand, the debate concerns the role of ruminants in building soil carbon, the role of grazing on albedo and in maintaining open cultural landscape and their biodiversity, the use of alternative metrics, the importance of animal proteins, and the loss of jobs for livestock farmers if such climate measure would be employed.

Grazing effects on soil carbon have been discussed in this chapter as having potential, but both in Norway and elsewhere the effects vary from one locality to the next. Potentials for carbon sequestration are highly dependent on the starting point of soil carbon, and are very likely not as large as suggested in some studies. Moreover, terrestrial carbon can be returned to the atmosphere on decadal timescales by a variety of mechanisms such as soil degradation, forest pest outbreaks and forest fires, and therefore requires careful consideration of policy frameworks to manage carbon storage, e.g. in forests (Gren & Aklillu, 2016). Open landscape and albedo effects have also been discussed and concluded that grazing must be intensive and lead to large vegetation changes in order to have a substantial effect. Interestingly, while building soil carbon through grazing and albedo effects are highlighted in the public debate as certain arguments for maintained or even increased cattle production, these measures or effects are not highlighted in the Norwegian agricultural- and national climate plan (KLD, 2021; Norges Bondelag, 2020), and are instead identified as uncertain factors where more research is needed.

This chapter has not addressed other environmental and biodiversity effects of grazing, but the international literature highlights that ruminant systems bring with them a range of environmental and societal concerns at the global aggregate level, including taking up a disproportionately large land area and water use for protein production (IPBES, 2019). Alternative ways of protein production – through non-ruminant meats, plant production or fish production – can produce the same amount of high quality protein using much less land and water (Poore & Nemecek, 2018; Ritchie, 2020). Biodiversity effects include the effects of grazing on rangeland biodiversity by removal of biomass, trampling and destruction of root systems, and replacement of wild grazers by livestock (Alkemade et al. 2013). Further, rangelands have been expanded at the cost of forested area either for grazing or for feed production, especially in regions in Africa and Brazil (IPBES, 2019; Lynch et al., 2020). On the other hand, local effects may be different from global effects, grazing may also stimulate biodiversity (Metera et al., 2010), and sustainable livestock management practices may also play a role in promoting biodiversity, safeguarding of native species, varieties, breeds and habitats, and ecological restoration (IPBES, 2019). The effects, in other words, are context dependent. While land released from livestock use cannot automatically be assumed to be used for something with lower climate and/or biodiversity consequences, potential non-agricultural uses of any spared land remain an important opportunity for climate change mitigation and other environmental aims. Indeed, most climate models suggest that globally we will require substantial additional land for carbon sequestration and bioenergy if we are to reach net-zero CO_2 emissions within a timeframe compatible with warming at 1.5 °C–2 °C above pre-industrial temperatures (Lynch et al., 2020).

We have not addressed the health aspects of eating meat in this chapter, as this is a complex and broad field including nuances in red-, white- and processed meat. While there are clear parallels between the health-meat and the climate-meat debate, a detailed discussion of health effects is beyond the scope of this chapter. However, we do wish to point out that while meat, especially lean meat, can make up a healthy contribution to a diet, meat as a protein and energy source is not a requirement for a healthy diet: while there are differences in quality and accessibility of plant- versus animal proteins, proteins and energy can also be had from white meat, fish, eggs, legumes or grains, all of which have lower overall emissions than ruminant-meat production (e.g. Poore & Nemecek, 2018; Ritchie, 2020). Both for health and for emission reasons, these high-income countries including the EU and Norway could reduce their intake of ruminant meats and substitute their proteins and energy with alternative sources such as white meat, fish, and legumes. Norway produces a vast amount of fish and has a larger plant-based production potential than it is currently using (Mittenzwei et al., 2017).

The climate metric GWP* has been discussed as an alternative to the commonly used GWP100, and the conclusion from that section is that the "big-picture" conclusion from most research on dietary sustainability does not change. Methane continues to rise globally, and climate targets such as the Paris Agreement 1.5 °C target cannot be reached without reductions in methane, which includes reductions in ruminant methane (e.g. Clark et al., 2020). Current knowledge and technologies for reducing enteric methane may play an important role in the future but are not well enough developed or of large enough effect to be upscaled and have significant emission reduction effects in living animals. The argument that near-stable methane emission (and number of cattle) would not contribute to increased warming when measured with the GWP* metric first ignores the CO_2 and nitrous oxide contribution of ruminants, secondly ignores the need to reduce methane emissions and not to keep emissions stable, and thirdly ignores the large – though one-off – cooling benefits of reductions in methane. The timing of such emission reductions can of course be discussed, although the behavioral change needed for consumption changes takes time and cannot be used as an "emergency brake" to keep warming from going over policy targets such as 1.5 °C. Finally, the argument that a stable number of cattle/methane emissions does not add to the climate problem is an example of "grandfathering", which is often regarded as being inequitable and hence strongly criticized (Rogelj & Schleussner, 2019). Under this principle, countries that historically have contributed to increasing methane pollution levels but now have stabilized these emissions are rewarded with the right to continue to emit similar shares in the future simply by virtue of having been first, while nations that are developing and increasing their meat consumption (and production) to levels comparable to high income countries are punished by not being allowed to further develop.

Finally, while loss of jobs for livestock farmers if such climate measure were employed is a very relevant concern, account should also be taken of the new jobs that would arise in alternative types of production, the fact that (at least in Norway) a reduction in number of livestock farms already takes place due to structural changes linked to low profitability, and the climate challenges that these farmers would face if emissions were not abated.

2.6.2.2 Measures and Opportunities for Emission Reductions

Plans for emission reductions through reduced ruminant meat production are largely absent both in the Norwegian national and agricultural sector climate plans. Instead, emission reductions are planned through a focus on (1) feed supplements to reduce methane in ruminants, improved feed quality, and breeding programs, and (2) carbon capture through use of biochar and carbon-capture crops as key points. Finally, this plan anticipates concessions in a relatively strict recently approved law and detailed regulations around a ban of wetland conversion, which in Norway stands for a high share of national emissions and is linked to farms expanding their farmland by draining wetlands to produce animal feed, typically grass for ruminants. An agreement between the farmers' union and the government states that the latter would actively work on reductions of red meat consumption and thus a reduced market demand. An important notion here is that while the current agricultural sector climate plan is planning to reduce emissions in its current production, it is not geared or preparing for a food system transition towards less meat production in line with global studies' recommendations (e.g. Clark et al., 2020).

Increasing profitability in agriculture (both in vegetable production but also in meat production) is important. A market balance in production-consumption helps maintain profitable food prices for farmers, while overproduction reduces prices and profitability for farmers (Stevanovic et al., 2017). If changes in demand are not matched by changes in production, Norway eventually will produce either too little or more than the national demand for meat. Producing too little means a need for increased import, which goes against the policy target of increasing national food security. Overproduction, if taking a regional view, could supply the international market. However, meat production in Norway is heavily subsidized, and seasonally fluctuating tariff barriers promote sales of Norwegian products over imported products within Norway. Since exports cannot be subsidized due to trade regulations, overproduction leads to a surplus of meat at low prices and low profitability for meat farmers, as there has been for many years. Secondly, a restructuring of the available productive land is necessary. Agricultural land use is currently sub-optimal, as good soils for production of food grains are increasingly being used for grass and cattle production (Tufte & Thuen, 2019). While total grain production nevertheless has increased due to improved production techniques, the sub-optimal use of land is largely caused by a higher profitability in grass and meat production which undermines increased food security in plant-based foods.

As the Norwegian national climate white paper states, changes in demand are critical and must happen prior to changes in production. Changing demand is one way to avoid a lock-in in unsustainable consumption patterns shaped by insufficient technological or other fixes on the production side. There is an increasing understanding and agreement about the potential of dietary change (Aleksandrowicz et al., 2016; Springmann et al., 2018) for several reasons including climate, health and environment, and implementation is possible at zero or even negative financial cost (Pettersen et al., 2017; Shindell et al., 2017). This makes for a compelling case for shifting diets, and above all for addressing consumption of ruminant meat. Many studies and policy documents (Departementene, 2016; Garnett et al., 2015; Poore & Nemecek, 2018; Tjärnemo & Södahl, 2015; Wellesley et al., 2015) point to the important role of industry, retailers and policy in such a shift, and that a staged approach involving these actors, multiple benefits and minimizing the shift from current dietary intakes are likely to make dietary change more realistic and achievable (Bailey et al., 2014; Horgan et al., 2016; Macdiarmid et al., 2016). Public awareness is increasingly understood to be insufficient: while attitudes towards reduced meat consumption seem to change (e.g. Bugge & Alfnes, 2018), consumption data show that a lack of incentives to reduce meat-production emissions is reflected in a continued increase in meat consumption (Helsedirektoratet, 2020). Retailers have a key role in procurement and potential to influence consumer behavior at the point of purchase through offering a greater range of alternative low-emission foods, through active product selection in procurement and sales, nudging and marketing (Garnett et al., 2015; Wellesley et al., 2015). However, a reluctance to transition away from ruminant meat is partly due to uncertainties related to the impacts and profitability (Bailey et al., 2014). Government incentives such as fiscal measures and procurement regulations (DIFI, 2018; Garnett et al., 2015; Wellesley et al., 2015) are crucial to promote changes in consumption and align government- and industry sector targets (Departementene, 2016).

2.7 Conclusions

Emissions from the global food system contribute substantially to global warming. Increasing meat production and consumption, and especially ruminant-meat production and consumption, plays a significant role in these emissions through increasing methane emissions, but also CO_2 and nitrous oxide emissions. There are also emission mitigating effects of some ruminant-management systems, such as grazing effects on albedo and the building of soil carbon. While important to include, the impacts of soil carbon and albedo on the carbon balance and warming effects of ruminant livestock production are spatially variable and uncertain in many locations. Soil carbon sequestration is not linear and sequestration rates decline with time. Moreover, sequestration can easily be reversed and is not necessarily a stable and permanent solution.

However, livestock production is not just a climate issue, but intertwined with traditions, production potentials, and many other local issues and food-related targets, as illustrated in this chapter through a case study on Norway. Consequently, policy and sector measures to reduce emissions from meat production in Norway have been limited to improved efficiency in production and carbon-sequestration measures – which do not include grazing. The key climate question is whether these proposed emission reductions with improved production practices are sufficient in combination with climate measures in other sectors to keep dangerous warming at bay.

The sum of emissions and sequestration in ruminant-meat production differ across countries. However, in the global quest to achieve the climate targets laid out in the Paris Agreement, it is difficult to see global mitigation that does not include reductions in meat production. Many studies find that strong reductions in consumption and production of especially high-emission meats are required to stay within sustainable limits for both climate and environment, and that failing to do so would lead to the Paris climate targets being out of reach, even if fossil fuel emissions were immediately halted.

This chapter acknowledges the complexity and the variety of issues linked to food production and consumption, especially ruminant meat. From a climate perspective however, our current understanding suggests that even when including improvements in meat production and accounting for potential climate-positive effects such as soil carbon uptake and albedo effects, ruminant meat still contributes substantially more to warming than alternative food production and consumption. This is the case in the Norwegian agricultural system. Many studies find that there remains a place for sustainable livestock production. But as shown in several studies, consumption-related measures are needed to avoid a lock-in of unsustainable consumption patterns which would be stimulated if we were only to concentrate on improving production measures. A two-pronged approach changing both diet and production, combined with improved efficiencies in production, are two complimentary ways that could contribute to reducing warming. Alone, these measures are not sufficient, and these must be combined with actions in other sectors as well.

The Paris climate targets require urgent and rapid mitigation measures. Globally, there are significant positive side effects from changes in production and consumption. The consequences of climate change beyond 1.5 °C–2 °C and the costs of damage and for adaptation may not be worth the gamble of insufficient action or leaving the task of emission reductions up to other sectors.

References

AgriAnalyse. (2019a). *Landbruksbarometeret 2019* (p. 53). AgriAnalyse. https://www.bondelaget.no/getfile.php/13907525-1557133354/MMA/Bilder%20NB/Arrangementer/Landbruksbarometeret%202019/Landbruksbarometeret%202019%20korr8.pdf

AgriAnalyse. (2019b). *Prosjektnotat januar 2019—Ammeku og produksjonsutvikling.*

Aleksandrowicz, L., Green, R., Joy, E. J., Smith, P., & Haines, A. (2016). The impacts of dietary change on greenhouse gas emissions, land use, water use, and health: A systematic review. *PLoS One, 11*(11), e0165797.

Alkemade, R., Reid, R. S., van den Berg, M., de Leeuw, J., & Jeuken, M. (2013). Assessing the impacts of livestock production on biodiversity in rangeland ecosystems. *Proceedings of the National Academy of Sciences, 110*(52), 20900–20905. https://doi.org/10.1073/pnas.1011013108

Allen, M., Cain, M., Lynch, J., & Frame, D. (2018). *Climate metrics for ruminant livestock.* Oxford Martin Programme on Climate Pollutants.

Amundsen, H., & Hermansen, E. A. (2020). Green transformation is a boundary object: An analysis of conceptualisation of transformation in Norwegian primary industries. *Environment and Planning E: Nature and Space, 251484862093433*. https://doi.org/10.1177/2514848620934337

Animalia. (2020). *Kjøttets tilstand 2020* (p. 152). Animalia.

Austrheim, G., Speed, J. D., Evju, M., Hester, A., Holand, Ø., Loe, L. E., Martinsen, V., Mobæk, R., Mulder, J., & Steen, H. (2016). Synergies and trade-offs between ecosystem services in an alpine ecosystem grazed by sheep–An experimental approach. *Basic and Applied Ecology, 17*(7), 596–608.

Bailey, R., Froggatt, A., & Wellesley, L. (2014). *Livestock–climate change's forgotten sector.* Chatham House.

Bright, R. (2019). Unpublished data.

Bright, R. M., Bogren, W., Bernier, P., & Astrup, R. (2016). Carbon-equivalent metrics for albedo changes in land management contexts: Relevance of the time dimension. *Ecological Applications, 26*(6), 1868–1880. https://doi.org/10.1890/15-1597.1

Bugge, A. B., & Alfnes, F. (2018). *Kjøttfrie spisevaner – hva tenker forbrukerne?* (nr. 14-2018; Oppdragsrapport, p. 93). Forbruksforskningsinstituttet SIFO, OsloMet – storbyuniversitetet.

Cain, M. (2018). *Guest post: A new way to assess 'global warming potential' of short-lived pollutants.* Carbon Brief. https://www.carbonbrief.org/guest-post-a-new-way-to-assess-global-warming-potential-of-short-lived-pollutants

Cao, L., & Caldeira, K. (2010). Atmospheric carbon dioxide removal: Long-term consequences and commitment. *Environmental Research Letters, 5*(2), 024011. https://doi.org/10.1088/1748-9326/5/2/024011

Caro, D., Davis, S. J., Bastianoni, S., & Caldeira, K. (2014). Global and regional trends in greenhouse gas emissions from livestock. *Climatic Change, 126*(1–2), 203–216. https://doi.org/10.1007/s10584-014-1197-x

Chandra, N., Patra, P. K., Bisht, J. S. H., Ito, A., Umezawa, T., Saigusa, N., Morimoto, S., Aoki, S., Janssens-Maenhout, G., Fujita, R., Takigawa, M., Watanabe, S., Saitoh, N., & Canadell, J. G. (2021). Emissions from the oil and gas sectors, coal mining and ruminant farming drive methane growth over the past three decades. *Journal of the Meteorological Society of Japan. Ser. II.* https://doi.org/10.2151/jmsj.2021-015

Chang, J., Peng, S., Ciais, P., Saunois, M., Dangal, S. R. S., Herrero, M., Havlík, P., Tian, H., & Bousquet, P. (2019). Revisiting enteric methane emissions from domestic ruminants and their δ13CCH4 source signature. *Nature Communications, 10*(1), 3420. https://doi.org/10.1038/s41467-019-11066-3

Clark, M., Hill, J., & Tilman, D. (2018). The diet, health, and environment Trilemma, 29.

Clark, M. A., Domingo, N. G. G., Colgan, K., Thakrar, S. K., Tilman, D., Lynch, J., Azevedo, I. L., & Hill, J. D. (2020). *Global food system emissions could preclude achieving the 1.5° and 2°C climate change targets*. 5.

CLEAR Center at UC Davis. (2020, July 7). *Why methane from cattle warms the climate differently than CO2 from fossil fuels*. CLEAR Center. https://clear.ucdavis.edu/explainers/why-methane-cattle-warms-climate-differently-co2-fossil-fuels

Climate Action Tracker. (no date). *Climate Action Tracker*. https://climateactiontracker.org/

Cole, J. R., & McCoskey, S. (2013). Does global meat consumption follow an environmental Kuznets curve? *Sustainability: Science, Practice and Policy, 9*(2), 26–36.

Davidson, A., & Wang, S. (2005). Spatiotemporal variations in land surface albedo across Canada from MODIS observations. *Canadian Journal of Remote Sensing, 31*(5), 377–390.

de Wit, H. A., Austnes, K., Hylen, G., & Dalsgaard, L. (2015). A carbon balance of Norway: Terrestrial and aquatic carbon fluxes. *Biogeochemistry, 123*(1–2), 147–173.

Departementene. (2016). *Kjente ressurser – uante muligheter Regjeringens bioøkonomistrategi* (No. 11/2016; p. 71). Nærings- og fiskeridepartementet.

DIFI. (2018). *Nye krav og kriterier til anskaffelser av bærekraftig mat og drikke*. https://www.anskaffelser.no/nyhet/2018/09/nye-krav-og-kriterier-til-anskaffelser-av-baerekraftig-mat-og-drikke-innspill-onskes

Een Thuen, A., & Tufte, T. (2019). *Kastrat – mer enn en nisjeproduksjon?* Bondebladet. https://www.bondebladet.no/kronikk/kastrat-mer-enn-en-nisjeproduksjon/

European Commission. (2020). *Farm to fork strategy. For a fair, healthy and environmentally-friendly food system*. European Union.

FAOSTAT. (2020). *Live Animals*. http://www.fao.org/faostat/en/?#data/QA

Farstad, F. M., Hermansen, E. A. T., van Oort, B., Grønlund, A., Mittenzwei, K., Brudevoll, K., & Sødal Grasbekk, B. (2020). *Forbudet mot nydyrking av myr: Bakgrunn, effekter og utfordringer* (CICERO Report 2020:11; CICERO Report) (p. 57). CICERO.

FFCC. (2021). *Farming for change. Mapping a route to 2030* (p. 52). Food, Farming and Countryside Commission (FFCC).

Garnett, T., Röös, E., & Little, D. (2015). *What is efficiency? And is it sustainable?* 48.

Garnett, T., Godde, C., Muller, A., Röös, E., Smith, P., de Boer, I., Zu Ermgassen, E., Herrero, M., van Middelaar, C., Schader, C., & van Zanten, H. (2017). Grazed and confused? *Grazed and confused?* READCUBE. http://scholar.google.com/scholar?q=GrazedandConfused?&btnG=&hl=en&num=20&as_sdt=0%2C22

Gerber, P. J. (2013). *Tackling climate change through livestock: A global assessment of emissions and mitigation opportunities* (Food and agriculture Organization of the United Nations, Ed.). Food and Agriculture Organization of the United Nations.

Global Carbon Project (GCP). (2021). *Global methane budget*. Global Carbon Project (GCP). https://www.globalcarbonproject.org/methanebudget/

Godfray, H. C. J., Aveyard, P., Garnett, T., Hall, J. W., Key, T. J., Lorimer, J., Pierrehumbert, R. T., Scarborough, P., Springmann, M., & Jebb, S. A. (2018). Meat consumption, health, and the environment. *FOOD SECURITY, 10*.

Gren, I.-M., & Aklillu, A. Z. (2016). Policy design for forest carbon sequestration: A review of the literature. *Forest Policy and Economics, 20*, 128–136.

Grønlund, A., Knoth de Zarruk, K., Rasse, D., Riley, H., Klakegg, O., & Nystuen, I. (2008). *Kunnskapsstatus for utslipp og binding av karbon i jordbruksjord* (No. 132/2008; Bioforsk Report) (p. 47). Bioforsk.

Gütschow, J., Jeffery, M. L., Gieseke, R., Gebel, R., Stevens, D., Krapp, M., & Rocha, M. (2016). The PRIMAP-hist national historical emissions time series. *Earth System Science Data, 8*, 571–603.

Gütschow, J., Jefferey, L., Gieseke, R., & Günther, A. (2019). *The PRIMAP-hist national historical emissions time series (1850–2017)* (Vol. 2.1). GFZ Data Services. https://doi.org/10.5880/PIK.2019.018

Hegrenes, A., & Asheim, L. J. (2006). *Verdi av fôr frå utmarksbeite og sysselsetting i beitebaserte næringar* (p. 34).

Helsedirektoratet. (2020). *Utviklingen i norsk kosthold 2019* (IS-2880) (p. 102). Helsedirektoratet.

Hillestad, M. E. (2019). *Beitemarka—Et ukjent karbonlager* (No. 5–2019; AgriAnalyse Rapport).

Hjelt, A. L., Jenssen, E., Hansen, Ø., Ystad, E., & Olsen, A. (2019). *Økonomien i grasbasert melk- og kjøttproduksjon i Nord-Norge* (No. 5, nr.140; NIBIO report) (p. 53). NIBIO.

Horgan, G. W., Perrin, A., Whybrow, S., & Macdiarmid, J. I. (2016). Achieving dietary recommendations and reducing greenhouse gas emissions: Modelling diets to minimise the change from current intakes. *International Journal of Behavioral Nutrition and Physical Activity, 13*(1). https://doi.org/10.1186/s12966-016-0370-1

IPBES. (2019). *The global assessment report on biodiversity and ecosystem services. Summary for policymakers* (p. 60). Intergovernmental Science-Policy Platform on Biodiversity and Ecosystem Services (IPBES).

IPCC. (2014). *Climate change 2014 synthesis report. Summary for policymakers* (IPCC AR5). IPCC.

IPCC. (2018). Summary for policymakers. In V. Masson-Delmotte, P. Zhai, H.-O. Pörtner, D. Roberts, J. Skea, P. R. Shukla, A. Pirani, W. Moufouma-Okia, C. Péan, R. Pidcock, S. Connors, J. B. R. Matthews, Y. Chen, X. Zhou, M. I. Gomis, E. Lonnoy, T. Maycock, M. Tignor, & T. Waterfield (Eds.). *Global warming of 1.5 °C. An IPCC special report on the impacts of global warming of 1.5 °C above pre-industrial levels and related global greenhouse gas emission pathways, in the context of strengthening the global response to the threat of climate change, sustainable development, and efforts to eradicate poverty*. IPCC. http://report.ipcc.ch/sr15/pdf/sr15_spm_final.pdf

IPCC. (2019a). *IPCC 2019 SPM SR climate change and land*.

IPCC. (2019b). *Summary for policymakers. IPCC special report on the ocean and cryosphere in a changing climate* (p. 36). IPCC.

KLD. (2017). *Meld. St. 41. Klimastrategi for 2030 – norsk omstilling i europeisk samarbeid*.

KLD. (2021). *Klimaplan for 2021–2030. Meld.St.13 (2020–2021). Melding til Stortinget* (Melding Til Stortinget Meld.St.13 (2020–2021); p. 212). Klima- og miljødepartement.

Klimautvalget. (2016). *Landbruk og klima – utredning fra arbeidsgruppe*.

Knutsen, H. (2018). *Utsyn over norsk landbruk. Tilstand og utviklingstrekk 2018* (Vol. 4). NIBIO.

LSB. (2020). *Livestock units*. https://www.lifestyleblock.co.nz/lifestyle-file/livestock-a-pets/the-basics/item/800-livestock-units

Lynch, J., Garnett, T., Persson, M., Röös, E., & Reisinger, A. (2020). *Methane and the sustainability of ruminant livestock*. 20.

Macdiarmid, J. I., Douglas, F., & Campbell, J. (2016). Eating like there's no tomorrow: Public awareness of the environmental impact of food and reluctance to eat less meat as part of a sustainable diet. Appetite, 96, 487–493. READCUBE. https://doi.org/10.1016/j.appet.2015.10.011.

Martinsen, V., Mulder, J., Austrheim, G., & Mysterud, A. (2011). Carbon storage in low-alpine grassland soils: Effects of different grazing intensities of sheep. *European Journal of Soil Science, 62*(6), 822–833.

Metera, E., Sakowski, T., Słoniewski, K., & Romanowicz, B. (2010). Grazing as a tool to maintain biodiversity of grassland – A review. *Animal Science Papers and Reports, 28*(4), 315–334.

Miljødirektoratet. (2020). *Klimakur 2030: Tiltak og virkemidler mot 2030* (M-1625 | 2020; p. 1191).

Mittenzwei, K., Milford, A. B., & Grønlund, A. (2017). *Status og potensial for økt produksjon og forbruk av vegetabilske matvarer i Norge*.

Monks, P. S., Archibald, A. T., Colette, A., Cooper, O., Coyle, M., Derwent, R., Fowler, D., Granier, C., Law, K. S., Mills, G. E., Stevenson, D. S., Tarasova, O., Thouret, V., von Schneidemesser, E., Sommariva, R., Wild, O., & Williams, M. L. (2015). Tropospheric ozone and its precursors from the urban to the global scale from air quality to short-lived climate forcer. *Atmospheric Chemistry and Physics, 15*(15), 8889–8973. https://doi.org/10.5194/acp-15-8889-2015

Ni, X., & Groffman, P. M. (2018). Declines in methane uptake in forest soils. *Proceedings of the National Academy of Sciences, 115*(34), 8587–8590. https://doi.org/10.1073/pnas.1807377115

Nisbet, E. G., Manning, M. R., Dlugokencky, E. J., Fisher, R. E., Lowry, D., Michel, S. E., Myhre, C. L., Platt, S. M., Allen, G., Bousquet, P., Brownlow, R., Cain, M., France, J. L., Hermansen, O., Hossaini, R., Jones, A. E., Levin, I., Manning, A. C., Myhre, G., et al. (2019). Very strong atmospheric methane growth in the four years 2014—2017: Implications for the Paris agreement. *American Geophysical Union, 38*. https://doi.org/10.1029/2018GB006009

Nordborg, M., & Röös, E. (2016). *Holistic management–a critical review of Allan Savory's grazing method. Holistic Management–a Critical Review of Allan Savory's Grazing Method.* READCUBE. http://publications.lib.chalmers.se/records/fulltext/244566/local_244566.pdf

Norges Bondelag. (2020). *Landbrukets klimaplan 2021–2030*. Norges Bondelag. https://www.bondelaget.no/tema/landbruketsklimaplan/landbrukets-klimaplan-pdf/

Norges Bondelag. (2021). *Bondelagets bærekraftstrategi for norsk matproduksjon 2021–2030*. https://www.digiblad.no/norges_bondelag/baerekraftstrategi_2021-2030/

Norwegian Environment Agency & Statistics Norway. (2021). *Klimagassutslipp fra jordbruk*. Miljøstatus. https://miljostatus.miljodirektoratet.no/tema/klima/norske-utslipp-av-klimagasser/klimagassutslipp-fra-jordbruk/

Nucci, D., Fatigoni, C., Amerio, A., Odone, A., & Gianfredi, V. (2020). Red and processed meat consumption and risk of depression: A systematic review and meta-analysis. *International Journal of Environmental Research and Public Health*, 20.

OECD. (2021). *Meat consumption*. https://doi.org/10.1787/fa290fd0-en. https://www.oecd-ilibrary.org/content/data/fa290fd0-en

Olofsson, J., Oksanen, L., Callaghan, T., Hulme, P. E., Oksanen, T., & Suominen, O. (2009). Herbivores inhibit climate-driven shrub expansion on the tundra: HERBIVORES INHIBIT SHRUB EXPANSION. *Global Change Biology, 15*(11), 2681–2693. https://doi.org/10.1111/j.1365-2486.2009.01935.x

Olofsson, J., te Beest, M., & Ericson, L. (2013). Complex biotic interactions drive long-term vegetation dynamics in a subarctic ecosystem. *Philosophical Transactions of the Royal Society B: Biological Sciences, 368*(1624), 20120486. https://doi.org/10.1098/rstb.2012.0486

Opio, C., Gerber, P., Mottet, A., Falcucci, A., Tempio, G., MacLeod, M., Vellinga, T., Henderson, B., & Steinfeld, H. (2013). Greenhouse gas emissions from ruminant supply chains – A global life cycle assessment. *Greenhouse gas emissions from ruminant supply chains – A global life cycle assessment.* READCUBE. http://scholar.google.com/scholar?q=Greenhouse gas emissions from ruminant supply chains – A global life cycle assessment.&btnG=&hl=en&num=20&as_sdt=0%2C22.

Pettersen, I., Grønlund, A., Elstad Stensgård, A., & Walland, F. (2017). *Klimatiltak i jordbruk og matsektoren. Kostnadsanalyse av fem tiltak* (NIBIO Report No. 3/2/2017, p. 99). NIBIO. https://docplayer.me/42544786-Klimatiltak-i-norsk-jordbruk-og-matsektor.html

Poore, J., & Nemecek, T. (2018). Reducing food's environmental impacts through producers and consumers. *Science, 360*(6392), 987–992. https://doi.org/10.1126/science.aaq0216

Rasse, D., Økland, I., Bárcena, T. G., Riley, H., Martinsen, V., Sturite, I., O'Toole, A., Øpstad, S., Cottis, T., & Budai, A. (2019). *Muligheter og utfordringer for økt karbonbinding i jordbruksjord* (NIBIO Report 5(36):2019; p. 94). NIBIO.

Reisinger, A., & Clark, H. (2018). How much do direct livestock emissions actually contribute to global warming? *Global Change Biology, 24*(4), 1749–1761. https://doi.org/10.1111/gcb.13975

Richi, E. B., Baumer, B., Conrad, B., Darioli, R., Schmid, A., & Keller, U. (2015). Health risks associated with meat consumption: A review of epidemiological studies. *International Journal for Vitamin and Nutrition Research, 9.*

Ridoutt, B. (2020). Climate neutral livestock production – A radiative forcing-based climate footprint approach. *Journal of Cleaner Production*, 125260. https://doi.org/10.1016/j.jclepro.2020.125260

Ritchie, H. (2019). *Food production is responsible for one-quarter of the world's greenhouse gas emissions*. Our World in Data. https://ourworldindata.org/food-ghg-emissions

Ritchie, H. (2020). *Environmental impacts of food production—Our world in data.* Published online at OurWorldInData.org. https://ourworldindata.org/environmental-impacts-of-food

Ritchie, H., & Roser, M. (2017, August 25). *Meat and dairy production*. Our World in Data. https://ourworldindata.org/meat-production

Rivedal, S. (2020). *Klimagassutslipp*. Nibio. https://www.nibio.no/tema/miljo/tiltaksveileder-for-landbruket/tiltak-mot-klimagassutslipp-fra-landbruket

Rogelj, J., & Schleussner, C.-F. (2019). Unintentional unfairness when applying new greenhouse gas emissions metrics at country level. *Environmental Research Letters, 14*(11), 114039. https://doi.org/10.1088/1748-9326/ab4928

Rogelj, J., Shindell, D., Jiang, K., et al. (2018). *Chapter 2: Mitigation pathways compatible with 1.5°C in the context of sustainable development* (IPCC SR1.5). IPCC.

Sans, P., & Combris, P. (2015). World meat consumption patterns: An overview of the last fifty years (1961–2011). *Meat Science, 109*, 106–111. READCUBE. https://doi.org/10.1016/j.meatsci.2015.05.012

Saunois, M., Stavert, A. R., Poulter, B., Bousquet, P., Canadell, J. G., Jackson, R. B., Raymond, P. A., Dlugokencky, E. J., Houweling, S., Patra, P. K., Ciais, P., Arora, V. K., Bastviken, D., Bergamaschi, P., Blake, D. R., Brailsford, G., Bruhwiler, L., Carlson, K. M., Carrol, M., et al. (2019). *The global methane budget 2000-2017* [preprint]. *Atmosphere – Atmospheric Chemistry and Physics*. https://doi.org/10.5194/essd-2019-128

Savory Institute. (2013). *Restoring the climate through capture and storage of soil carbon using holistic planned grazing* (p. 20). Savory Institute. https://savory.global/wp-content/uploads/2017/02/restoring-the-climate.pdf

Schmitz, O. J., Raymond, P. A., Estes, J. A., Kurz, W. A., Holtgrieve, G. W., Ritchie, M. E., Schindler, D. E., Spivak, A. C., Wilson, R. W., & Bradford, M. A. (2014). Animating the carbon cycle. *Ecosystems, 17*(2), 344–359.

Schwaab, J., Bavay, M., Davin, E., Hagedorn, F., Hüsler, F., Lehning, M., Schneebeli, M., Thürig, E., & Bebi, P. (2015). Carbon storage versus albedo change: Radiative forcing of forest expansion in temperate mountainous regions of Switzerland. *Biogeosciences, 12*(2), 467–487. https://doi.org/10.5194/bg-12-467-2015

Shindell, D. T., Fuglestvedt, J. S., & Collins, W. J. (2017). The social cost of methane: Theory and applications. *Faraday Discussions, 200*, 429–451. https://doi.org/10.1039/C7FD00009J

Sillmann, J., Aunan, K., Emberson, L., Büker, P., van Oort, B., O'Neill, C., Otero, N., Pandey, D., & Brisebois, A. (2021). Combined impacts of climate and air pollution on human health and agriculture. *Environmental Research Letters, 16*(9). https://doi.org/10.1088/1748-9326/ac1df8

Skagen, K. (2020). *Bærekraftig matproduksjon En diskursiv tilnærming til koordinering* [Master thesis]. University of Oslo.

Speed, J. D., Martinsen, V., Mysterud, A., Mulder, J., Holand, Ø., & Austrheim, G. (2014). Long-term increase in aboveground carbon stocks following exclusion of grazers and forest establishment in an alpine ecosystem. *Ecosystems, 17*(7), 1138–1150.

Springmann, M., Clark, M., Mason-D'Croz, D., Wiebe, K., Bodirsky, B. L., Lassaletta, L., de Vries, W., Vermeulen, S. J., Herrero, M., Carlson, K. M., Jonell, M., Troell, M., DeClerck, F., Gordon, L. J., Zurayk, R., Scarborough, P., Rayner, M., Loken, B., Fanzo, J., et al. (2018). Options for keeping the food system within environmental limits. *Nature, 562*, 519–525. https://doi.org/10.1038/s41586-018-0594-0

Statistics Norway. (1999). *Classification of livestock—Statistics Norway*. https://www.ssb.no/en/klass/klassifikasjoner/29

Statistics Norway. (2021a). *Emission to air. Table 08940: Greenhouse gases, by source, energy product and pollutant 1990—2020*. PX-Web SSB. http://www.ssb.no/en/statbanken/statbank/table/08940/

Statistics Norway. (2021b). *Holdings, agricultural area and livestock. Table 12660: Livestock grazing on outfield pastures (M) 1995—2020*. PX-Web SSB. http://www.ssb.no/en/statbanken/statbank/table/12660/

Statistics Norway. (2021c). *Livestock husbandry. Table 03710: Domestic animals per 1 March, by type of herd 1990—2021*. PX-Web SSB. http://www.ssb.no/en/statbanken/statbank/table/03710/

Stevanovic, M., Popp, A., Bodirsky, B. L., Humpeno, F., Mu, C., Kreidenweis, U., Rolinski, S., Biewald, A., & Wang, X. (2017). Mitigation strategies for greenhouse gas emissions from agriculture and land-use change: Consequences for food prices. *Environmental Science & Technology, 10*.

te Beest, M., Sitters, J., Ménard, C. B., & Olofsson, J. (2016). Reindeer grazing increases summer albedo by reducing shrub abundance in Arctic tundra. *Environmental Research Letters, 11*(12), 125013. https://doi.org/10.1088/1748-9326/aa5128

Tjärnemo, H., & Södahl, L. (2015). Swedish food retailers promoting climate smarter food choices—Trapped between visions and reality? *Journal of Retailing and Consumer Services, 24*, 130–139. https://doi.org/10.1016/j.jretconser.2014.12.007

Totland, T. H., Meln, B. K., Lundberg-Hallén, N., Helland-Kigen, K. M., Lund-Blix, N. A., Myhre, J. B., Johansen, A. M. W., Løken, E. B., & Andersen, L. F. (2012). *Norkost 3 En landsomfattende kostholdsundersøkelse blant menn og kvinner i Norge i alderen 18-70 år, 2010-11* (p. 70). Helsedirektoratet.

Tufte, T., & Thuen, A. E. (2019). *Arealutvikling og arealbruk—Korn, gras og drøvtyggere* (AgriAnalyse Rapport No. 12–2019; p. 59). AgriAnalyse.

van den Pol, A., Chabbi, A., de Vliegher, A., Hennessy, D., Hutchings, N., & Klumpp, K. (2018). *Grazing for carbon: End report*. EIP-AGRI.

Weindl, I., Popp, A., Bodirsky, B. L., Rolinski, S., Lotze-Campen, H., Biewald, A., Humpenöder, F., Dietrich, J. P., & Stevanović, M. (2017). Livestock and human use of land: Productivity trends and dietary choices as drivers of future land and carbon dynamics. *Global and Planetary Change, 159*, 1–10. https://doi.org/10.1016/j.gloplacha.2017.10.002

Wellesley, L., Happer, C., & Froggatt, A. (2015). Changing climate, changing diets. Pathways to lower meat consumption. 8.

Chapter 3
The Limits of Vegetarianism

Diane Mayerfeld

Abstract Limiting meat consumption can and must play an important role in reducing global greenhouse gas emissions, and Chaps. 8 and 9 will look at ways to approach that challenge. However, the importance and effectiveness of reducing meat consumption has often been exaggerated and distorted. In the United States, Canada, and the European Union greenhouse gas emissions from transportation far exceed emissions from the whole agricultural sector, as do emissions from other sectors such as industry and energy production. In addition, livestock plays a critical role in organic and sustainable agricultural systems, and significant declines in beef consumption in the U.S. have not been correlated with a reduction in agriculture's greenhouse gas emissions. When the only two options discussed are continuing current trends in livestock production or eliminating livestock agriculture, neither farmers nor policymakers are likely to make constructive changes.

Keywords Beef · Climate change · Diet · Greenhouse gas emissions · Livestock · Sustainable agriculture

Limiting meat consumption can and must play an important part in reducing global greenhouse gas emissions, and Chaps. 8 and 9 will look at ways to approach that challenge. However, the importance and effectiveness of reducing meat consumption has often been exaggerated and distorted. This chapter will explore some of those distortions and will also lay out some of the reasons why it is important to retain livestock agriculture. When the only two options discussed are continuing current trends in livestock agriculture on one hand, or eliminating livestock agriculture altogether on the other, neither farmers nor policymakers are likely to make constructive changes.

D. Mayerfeld (✉)
Division of Extension, Agriculture Institute, University of Wisconsin,
Madison, WI, USA
e-mail: dbmayerfeld@wisc.edu

D. Mayerfeld (ed.), *Our Carbon Hoofprint*, Food and Health,
https://doi.org/10.1007/978-3-031-09023-3_3

3.1 Excess Claims

Mark Twain famously quipped in *Chapters from My Autobiography* that there are three kinds of lies: lies, damn lies, and statistics. Too often, the statistics describing the contribution of livestock agriculture to climate change are taken out of context, and that lack of context has given many people in wealthy nations the false impression that reducing their meat consumption is the most important action they can take (and the only action they need to take) to address climate change.

Take this e-mail a student organization sent to all the students at my university a few years ago:

> Do you care about protecting our environment? Do you think we should continue wasting resources and polluting our environment? Here's some unsettling facts you should know:
>
> - Raising farm animals produces more greenhouse gas emissions than all planes, trains, and automobiles combined.
> - Nearly 80% of land deforested in the Amazon is used for raising cows.
> - To produce one pound of animal protein vs. one pound of soy protein, it takes about 12 times as much land, 13 times as much fossil fuel, and 15 times as much water. You could save more clean water by not eating one pound of beef than not showering for six months!
> - Switching to a diet free of meat, dairy, and eggs saves more carbon emissions than driving a Prius – 50% more!
> - Raising animals for food now uses a staggering 30% of the Earth's land mass.
> - According to the UN, eating animal products is one of "the most significant contributors to the most serious environmental problems, at every scale from local to global."
>
> …
>
> "If anyone wants to save the planet, all they have to do is just stop eating meat. That's the single most important thing you could do. It's staggering when you think about it. Vegetarianism takes care of so many things in one shot: ecology, famine, cruelty." -Sir Paul McCartney.

Where does this argument come from, and what is wrong with it? Let's look at a few of its claims in more detail.

The statement that "Raising farm animals produces more greenhouse gas emissions than all planes, trains, and automobiles combined" almost certainly comes from the executive summary of *Livestock's Long Shadow*.[1] However, there are a couple of issues with the way this claim is used here. First, by 2018 when this email was sent, an updated FAO report had revised its estimates of livestock greenhouse gas emissions and concluded that the livestock sector's emissions accounted for about 14.5% of global anthropogenic emissions, less than transportation's growing contribution.[2] More importantly, though, this email was addressed to students in the United States, where transportation contributes more than 27% of total greenhouse gas emissions, while all of agriculture accounts for between 10% and 12% of

[1] Steinfeld et al. (2006).

[2] Gerber et al. (2013).

U.S. greenhouse gas emissions, and about two thirds of those total agricultural emissions can be attributed to livestock (Fig. 3.1). So in the U.S., where these students live, greenhouse gas emissions from all planes, trains, and cars combined are more than three times greater than the emissions from livestock production.

Americans eat a lot of meat compared to most other cultures.[3] So how is it possible that livestock contributes so little to U.S. greenhouse gas emissions, even when U.S. meat exports are greater than our imports? The answer has three main parts. First, as the next chapter will explain, U.S. and European livestock production is relatively efficient and emits less CO_2 and methane per unit of meat than many other countries. Second, Americans live in large houses that take a lot of energy to heat, cool, and light; and Americans also drive much more than most of the world. On top of that, we do our driving in inefficient vehicles. And third, livestock agriculture in the U.S. is not usually associated with deforestation. As we saw in Chap. 1, carbon emissions from tropical deforestation account for around 30% of the world's greenhouse gas emissions from livestock production, but almost all that deforestation is occurring in South America. There is some land use change in U.S. agriculture, but it is relatively small and swings back and forth with grain prices and farm policy. For the first 300 years after European settlement U.S. agriculture expanded into

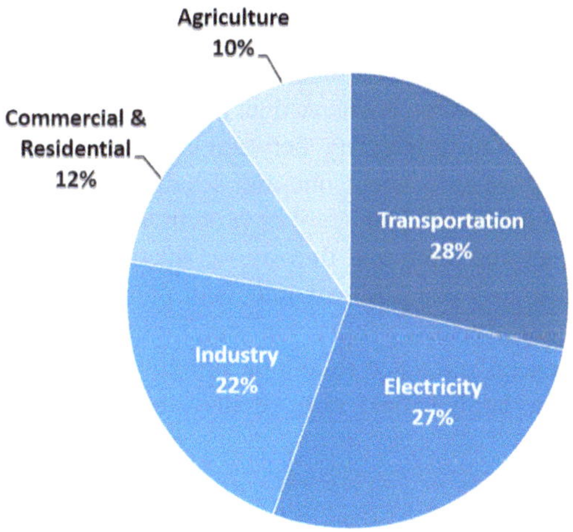

Fig. 3.1 Total U.S. greenhouse gas emissions by economic sector in 2018. (U.S. Environmental Protection Agency 2020)

[3] In 2019 per capita meat consumption (including beef, pork, lamb, and poultry) in the U.S. was estimated at just over 100 kg per year. Australia, Argentina, and Israel were not far behind, with 88–90 kg per capita meat consumption. Norwegians ate about 56 kg, the world average was 34 kg, and per capita meat consumption in India, Ethiopia, and Nigeria was less than 5 kg (OECD, 2021).

forest and prairie, undoubtedly releasing a great deal of CO_2 in the process, but for the last 70 years the total amount of land in agriculture in the U.S. has remained relatively constant, and forest land has increased slightly.[4]

The source for the statement that "nearly 80% of land deforested in the Amazon is used for raising cows" is probably an article by Nepstad et al. in which the authors write that cattle ranching is "associated with four-fifths of Amazon deforestation."[5] Note that the term "associated with," does not necessarily attribute causation. This and other articles about tropical forest loss discuss a complex mix of factors affecting Amazonian deforestation, including domestic Brazilian policy, international trade regulation, currency values, and ethanol demand, as well as the market for meat.[6] While rising global demand for meat is probably an important contributor, it is far from the only driver of tropical deforestation. Without context, however, readers of the email above are meant to infer that beef consumption causes 80% of tropical deforestation and that therefore a reduction in beef consumption would result in a corresponding reduction in deforestation – or possibly even in land abandonment and reforestation.

The statement about how many resources a pound of animal protein takes to produce compared to a pound of soybean protein raises a dizzying array of questions. I will touch on just a few of them here.

First, what is meant by a pound of animal protein? Many authors have addressed the question of how much land, water, and energy it takes to produce meat compared to grain or pulses such as soybeans or lentils. While their estimates vary depending on their assumptions and models, all find that on average raising meat is more resource intensive than growing grain and pulses, and all find great differences in resource intensity between livestock types. Typically, these studies find that chicken uses about one and a half to twice as much land, water, and energy per pound, calorie, or gram of protein as pulses such as soybeans, while estimates for beef and lamb find they are on average somewhere between 10 to 50 times more resource intensive per unit than plant-based alternatives. Pork is more resource-intensive than poultry, but much less so than beef.[7] Given this wide variation, the general reference to animal protein is misleading, particularly since the numbers cited apply to beef, rather than to an average of livestock products. At the same time,

---

[4] U.S. Department of Agriculture NRCS (2013), Alig et al. (2010), Oswalt and Smith (2014).

[5] Nepstad et al. (2009).

[6] Nepstad et al. (2009), Gibbs et al. (2016), Katz-Rosene (2020).

[7] Eshel et al. (2014), Gidon Eshel et al. (2018), Pimentel and Pimentel (2003), Harwatt et al. (2017).

soybeans are particularly high in both protein and calories; they contain more than twice as much per gram as other plant foods such as kidney beans and pinto beans.[8]

Second, what is meant by water use? Pimentel was one of the first scholars to compare water used in industrial meat production to that needed to raise plant-based foods. He noted that water used directly for livestock production (to allow animals to drink and in processing livestock products) is relatively low, and accounts for approximately 1.3% of agriculture's water use. However, if one counts all the water needed to raise the feed eaten by livestock, then livestock water use is high. Including the water needed to grow their feed, Pimentel estimates that it takes almost twice as much water to produce a kilogram of chicken as to produce a kilo of soybeans: 3500 liters for chicken, compared to 2000 liters for soybeans. And he estimates it can take 100,000 liters of water to produce a kilo of beef – 50 times as much as needed to raise a kilo of soybeans.[9] Again, by this measure the resource intensity of poultry is considerably lower than that of beef.

However, most of the water that is "used" for beef production is rainfall on pasture or range. On well-managed pasture the presence of the cattle does not affect the water cycle or quality – the rain soaks into the soil, some of it goes to recharge rivers or the aquifer, and the rest is taken up by grass and transpired back to the atmosphere, just as if no cattle were present. On the other hand, if the land is overgrazed, over-fertilized, or cattle access to watercourses is mismanaged, then the impact on water quality and hydrology can be severe. Likewise, irrigation of pasture or forage crops can cause adverse changes to local (or distant) hydrology, as well as greenhouse gas emissions from the energy needed to move the water.[10]

Scientists and water managers distinguish between the different ways water can be used to grow crops and label them as green, blue, and grey water use.[11] Green water is rainfall, blue water is irrigation water, and gray water is a measure of the amount of water that is polluted as a result of its use. Total water use per cow might be the same, but if one cow gets all her feed from well-managed rainfed pasture her impact on water quantity and quality is very different from that of a cow that is fed

[8] Bohrer (2017), Harwatt et al. (2017). When thinking about actual dietary implementation the picture gets more complicated still. Plain dried soybeans have about twice as much protein per gram as beef or chicken, but people don't eat plain dried soybeans. Instead they eat tofu (or other processed soybean products), which contains roughly half as much protein per 100-gram serving as most meats. It is probably not coincidental that articles on the advantages of plant-based proteins use high-protein beans such as plain soybeans or black beans, rather than other familiar plant foods like pinto beans or hummus. Comparisons of the protein and other nutritional content of legumes and meat can be confusing, particularly when they are based on raw product and not on food serving. While meat typically loses a very small amount of weight with cooking as some of its water content evaporates, in most cases beans, peas, and lentils will more than triple their weight in cooking as they absorb water (Sweers, 2013).

[9] Pimentel et al. (1997).

[10] Hoekstra et al. (2012).

[11] Hoekstra and Mekonnen (2012). See also the Grace Communications Water footprint calculator at https://www.watercalculator.org/footprint/water-footprint-beef-industrial-pasture/

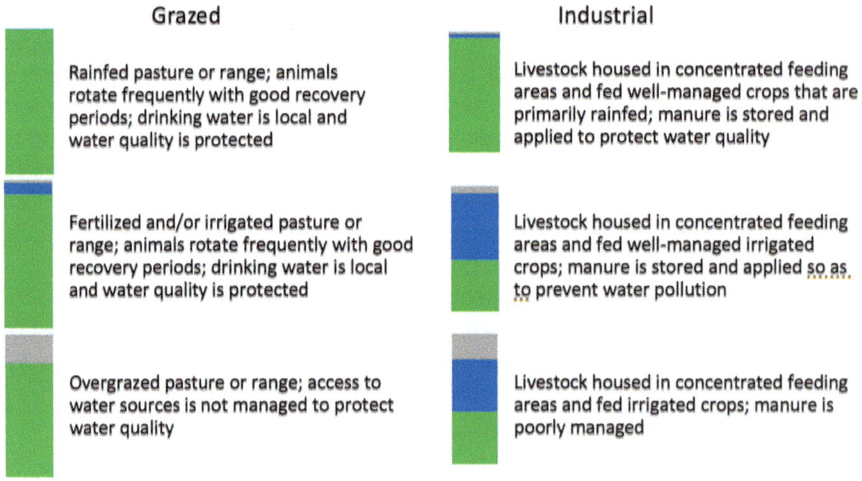

Fig. 3.2 Hypothetical green, blue, and grey water footprints of beef produced under different scenarios

irrigated crops, especially if the water comes from far away or is contaminated with nitrates or pesticides applied to cropland (Fig. 3.2).

When Mekonnen and Hoekstra modeled green, blue, and grey water use for livestock products by animal type and country they found wide variations between livestock types, countries, and production types. For example, in India their model suggests grass-based beef or buffalo production would use 25,913 m^3/ton green water, 242 m^3/ton blue water, and no grey water, while industrial beef or buffalo there would use about half as much green water but six times as much blue water and 866 m^3/ton grey water. However, industrial beef production in the U.S. and Netherlands would use less water per ton in all categories than grass-fed beef, and much less water than Indian beef (because cattle in Europe and the U.S. grow faster and so need less water and feed over the course of their lives). The least water-intensive meat in their model was chicken in the Netherlands under a mixed grazing and industrial system, which used 1509 m^3/ton green water, 76 m^3/ton blue water, and 161 m^3/ton grey water.[12]

Of course water use and impacts can also vary widely within a country as well as between countries. Ridoutt et al. compared water used for beef production on 6 farms in Australia and found wide variation in the amount of water used for irrigation, drinking, and other farm operations – from 25 liters per kg live weight to 234 l/kg. Taking into account water availability in the area, they found a more than 60-fold difference in water footprint between the most and least water-intensive operation.[13] The U.S. also has wide regional variation in blue water use to raise beef, from an

[12] Hoekstra and Mekonnen (2012).

[13] Ridoutt et al. (2012).

average of 192 l/kg carcass weight in the northeastern U.S. to 6187 l/kg carcass weight in the northwestern part of the country, which relies heavily on irrigation to grow alfalfa and other feed crops.[14] Blue water footprints can also vary significantly within each region. Livestock production can negatively affect water quality and availability, but those impacts differ widely depending both on farming practices and on farm location.[15]

This explanation of different types of water use brings us to the statement that "You could save more clean water by not eating one pound of beef than not showering for six months!" That claim could be true if the person takes short showers and if most of the animal's feed was irrigated, or if the farm caused significant water pollution. But although pastures need rain to grow forage, the presence of cattle on those pastures doesn't affect water in the same way that using water for a shower does. In a typical American or European household shower water first requires energy to pump it to a treatment plant, where it will be treated with chlorine or other chemicals or processes. Older public water systems can have significant leakage in water delivery, so each liter that reaches the home might reflect up to one and a half liters of water that was withdrawn from the water source and treated. Before the shower, the water is heated, and after the shower the now soapy water is mixed with sewage, treated again, and then discharged into a different place. Thus, water used in a shower typically has a greater environmental impact than rain on a pasture or range.

This is a long digression on water impacts of livestock for a book focused on greenhouse gas emissions. It is relevant, though, because both scholarly and popular articles about the climate footprint of meat often also address the impacts livestock has on water and land use. And indeed, we cannot and should not let climate change be the only environmental impact we consider. In addition, the argument that eating meat is the main cause of climate change often relies on very similar simplifications and distortions of statistics as this claim about water use.

Which brings us to the statement that "Switching to a diet free of meat, dairy, and eggs saves more carbon emissions[16] than driving a Prius – 50% more!" The precise percentage gives the impression that this statement is based on a strong body of research, but if you look more carefully it becomes clear that the comparison makes no sense. How much meat – and what kind of meat – does this hypothetical diet contain before the person stops eating livestock products? What does the person eat instead? If the person substitutes soybeans for all livestock products the greenhouse gas reductions will be greater than if the person increases their consumption of palm oil, coconut cream, and exotic fruits to compensate for meat, dairy, and eggs. And

[14] Rotz et al. (2019).

[15] Rose et al. (2020).

[16] In fact, the *carbon* impacts of livestock raised in the U.S. are not very high – it is the methane and nitrous oxide emissions from enteric fermentation and manure that make up most of the greenhouse gas emissions from livestock in the U.S. European livestock are more likely to be fed soybeans from Brazil, which can be associated with significant carbon emissions if they are grown on recently deforested land.

the second half of the statement is even more lacking in specificity. Driving a fuel-efficient car does not in itself reduce carbon emissions. Presumably the assumption is that the person is switching from an inefficient car to the Prius, though the statement does not actually say that, nor does it state how many kilometers the person supposedly drives.

So let's look at a U.S. resident who drives an average amount and eats an average U.S. diet, high in livestock products. All of agriculture accounts for approximately 10% of U.S. greenhouse gas emissions. Livestock, including livestock feed production, may contribute up to 80% of those emissions, so by eliminating all livestock products from their diet *and not substituting any other foods* this average American could theoretically reduce their total greenhouse gas footprint by 8%.

Now assume that instead that person continues to eat meat but switches from a car with average fuel efficiency to a Prius. Transportation currently accounts for 28% of U.S. greenhouse gas emissions, and light duty vehicles (private cars) contribute 59% of those greenhouse gas emissions, or about 16.5% of all U.S. greenhouse gas emissions. In 2017 the average new light duty vehicle in the U.S. had a fuel efficiency of 22.3 mpg or 9.5 km/l, while a Prius had a fuel efficiency of about 50 mpg or 21 km/l.[17] By switching from an average car to a Prius this person would reduce their total greenhouse gas emissions by about 9%, slightly more than by eliminating all livestock products from their diet. If the person switches from a conventional sport utility vehicle or truck to a Prius, then their emissions reduction would be greater than 9%.

In contrast, for an average European switching to a Prius would not reduce greenhouse gas emissions much at all, simply because by 2015 the average private car in Europe was more than twice as fuel efficient as the average U.S. car, and nearly as efficient as a Prius.[18] The average European would have to cut back on plane trips or number of kilometers driven to reduce their transportation footprint. Context and details matter.

3.2 The Impact of These Distortions

But does it really matter if these statements are a little sloppy? The fact that for Americans and Europeans electricity use and transportation are larger sources of greenhouse gas emissions than diet does not mean that we should ignore the impact of our agriculture and diet. It is true that on average meat, especially beef and lamb, causes more greenhouse gas emissions per unit than plant substitutes such as soybeans. Moreover, many people eat more meat than the recommended nutritional

---

[17] Bureau of Transportation Statistics (n.d.), U.S. Department of Energy and U.S. Environmental Protection Agency (n.d.).

[18] Hu and Chen (2016). However, European fuel economy statistics are based on ideal laboratory conditions rather than actual mileage, so switching to a Prius may be more beneficial than the official statistics would suggest (Tietge et al., 2017).

guidelines, which may contribute to diet-related health issues such as heart disease.[19] If exaggerating the benefits of a vegan diet will lead people to eat less meat what is wrong with that?

There are two serious problems with this type of exaggerated claim. The first is that inaccurate information can lead well-intentioned individuals to make ineffective choices and advocate for ineffective policies. If consumers in wealthy nations reduce the amount of meat and dairy they eat but don't rein in fossil fuel use for transportation and electricity they will have little impact on overall greenhouse gas emissions. It is even possible that if consumers think that reducing meat consumption will reduce the greenhouse gas emissions associated with their diet, regardless of other dietary adjustments, they could actually increase their dietary emissions. While most models suggest that reducing the amount of meat in north American and European diets would reduce greenhouse gas emissions, one analysis found that if all Americans reduced their meat consumption and increased their fruit, vegetable, and dairy consumption to meet current U.S. nutritional guidelines, their dietary greenhouse gas emissions and water footprint would go up, not down.[20] Grains and pulses have consistently low greenhouse gas emissions per unit of protein and per calorie, but the greenhouse gas emissions and water use associated with fruits and vegetables are highly variable, and may in some situations exceed those of chicken.[21]

An even greater problem with this exaggerated rhetoric of livestock's impact is that it discourages good-faith negotiation and cooperation. The livestock industry sees that many of the claims about the impact of meat and dairy consumption on the climate are exaggerated. They also see that advocates for reducing the impact of livestock frequently base their comparisons on a complete elimination of animal products, rather than a reduction in consumption and improvement in production practices. Think about "Switching to a diet free of meat, dairy, and eggs saves more carbon emissions than driving a Prius – 50% more!" The comparison is between a complete elimination of all livestock products, and a – change in car brand. The default to a vegan standard leads farmers and others who work with animal agriculture to view the movement to connect livestock to climate change as an existential threat.

---

[19] Bentley (2017).

[20] Tom et al. (2016).

[21] Here again, assumptions and framing matter. With the exception of lettuce grown in heated greenhouses, a review of greenhouse gas emissions for fruits, vegetables, grains, and meats concluded that emissions per kilogram of product were higher for meat than for fruits and vegetables (Clune et al., 2017). However, this review excluded air transport from their calculations; they compared boneless meat to whole vegetables, and for many of their food categories the deviation from the mean was greater than 50%, indicating that at least some of the studies reviewed found high emissions for plant foods or low emissions for meats. Since many fruits and vegetables have high water content, comparing emissions on the basis of weight rather than protein or energy will also tend to favor plant products. Finally, as Helander et al. (2021) note, fruits and vegetables typically are associated with high amounts of spoilage and food waste at every level from farm to table, further complicating comparisons.

A few years ago the organization I work for, University of Wisconsin Cooperative Extension, conducted an internal climate awareness exercise. It was designed to be a fun way for people to learn about how they could reduce their climate footprints. Employees could form teams and get points for activities such as buying energy-efficient appliances, car-pooling, recycling, and reducing meat consumption. Our Cooperative Extension program has four branches: one focused on working with farmers, one on helping communities plan for economic development and environmental protection, one on helping families with nutrition and financial management, and one focused on youth leadership development. Several of my agriculture colleagues quickly pushed back against the part of the game that awarded players points for reducing meat consumption. Colleagues working in the community development and nutrition programs responded with accusations that agriculture didn't respect science or care about climate change. I was able to bring some of the more vocal people on both sides together for a series of conversations about the role of animal agriculture in climate change.

To me, the most revealing moment came in the final conversation when one of the community development colleagues advocating for a reduction in meat consumption said in exasperation that of course she didn't want to get rid of animal agriculture altogether, and all those advocating for a reduction in animal agriculture agreed. Suddenly, the tone of the conversation changed from wary and even hostile to cooperative. The (Prius-driving) dairy agent spoke about his work to get local food processors to sell or donate food waste to local dairies, which would reduce greenhouse gas emissions from landfilling that waste, from growing more corn silage, and from enteric emissions. The (vegetarian) community development agents expressed their appreciation for sustainable agriculture, including some livestock production. They were still far apart on the specifics of how many cows are too many and the relative importance of diet and other activities for greenhouse gas emissions, but negotiation, cooperation, and compromise seemed possible, where they had not before.[22]

Not surprisingly, much of the livestock industry has responded to exaggerated rhetoric about the role of animal agriculture in driving climate change (and the implied threat to their livelihood) by simplifying and cherry-picking their own data, rather than by seeking ways to reduce their emissions. At least among my agriculture colleagues in Wisconsin, even those who are concerned about climate change and are working to improve production practices are reluctant to discuss a reduction in livestock numbers. I suspect this reluctance comes because in an adversarial bargaining process where veganism is the stated goal of one side the natural response is to stake out an equally extreme opposite goal.

In both North America and Europe agriculture has far more political power than one would expect based on the number of farmers. In many ways climate change has hurt farmers more than those whose livelihood doesn't depend directly on the weather, yet at least in the U.S. most farm organizations have resisted policies and

[22] Mayerfeld et al. (2016).

laws designed to reduce greenhouse gas emissions. It's not that they can't tell that the climate is changing. The state of Wisconsin has been less affected by climate change than most of the U.S. in that our summer days have not (yet) become hotter or drier. Still, Wisconsin farmers know the average frost-free season is 2–3 weeks longer than it was 50 years ago, and they know that rainfall, especially in spring and fall, is much higher and more intense than it used to be, disrupting critical planting and harvest seasons. The resistance to climate change action stems in large part, I believe, from a fear that climate activists expect agriculture to eliminate its greenhouse gas emissions without regard for the feasibility of that goal or the economic impact on farmers.[23]

Sadly, it is not just naïve undergraduates who make misleading statements and advocate for elimination rather than change of livestock production. Some leading academics actively promote these views. For example, Joseph Poore, a researcher at the University of Oxford told *The Guardian* newspaper in 2018:

> A vegan diet is probably the single biggest way to reduce your impact on planet Earth, not just greenhouse gases, but global acidification, eutrophication, land use and water use. It is far bigger than cutting down on your flights or buying an electric car.[24]

Why is the idea that cutting meat consumption is the most important action to combat climate change so widely promoted?

One reason is the idea that it is easy for individuals to change their diet. Other individual behaviors that reduce greenhouse gas emissions can cost a lot of money up front, like putting solar panels on the roof or buying more energy-efficient appliances. You also have to own your home to make changes of that kind. Still other climate-friendly actions may save money right away but seem like significant sacrifices, such as moving to a smaller house or giving up a plane trip. But surely giving up one serving of meat a week is an easy change to make?

There is some truth to this idea, and as we will explore further in Chaps. 8 and 9, reducing consumption of livestock products can help lessen agriculture's climate hoofprint, *as long as that reduction in consumption is linked to specific changes in farming practices that will reduce net greenhouse gas emissions from agriculture*. But the situation is a little more complicated. While it is easy for most people who have an ample diet to give up one or two servings of meat or dairy per week, it does not follow that it is easy for most people to give up animal foods altogether, and claims about the climate impact of giving up just a little meat should not be exaggerated. Moreover, the decision to not eat meat or dairy has to be made over and over, and it is just as easy to go back to eating more meat as to decrease consumption.[25] In contrast, those initially more difficult choices like buying energy-efficient appliances or choosing to live close to your workplace don't require continued effort to sustain lower greenhouse gas emissions. And they are just as difficult to reverse as they are to take in the first place.

---

[23] Doll et al. (2017).

[24] Carrington (2018).

[25] Asher et al. (2014), Haverstock and Forgays (2012).

3.3 Unintended Consequences

An even greater problem with relying on changes in consumer demand for meat is that reduced demand may not necessarily produce the desired reductions in greenhouse gas emissions. One of the simplistic assumptions of the push to reduce meat consumption is the idea that as demand drops farmers will stop raising livestock, and will not replace their animals with another equally or more damaging agricultural activity.[26] But farmers have to make a living, and when the demand for their product drops, most do not have the luxury of simply letting their land revert to forest or prairie. So they seek other ways to get money from their land, and some of those alternatives can emit more greenhouse gases than livestock. A look at recent trends in U.S. diets and agricultural emissions illustrates the reality that a drop in demand for meat may not have the predicted results.

As Figs. 3.3 and 3.4 illustrate, both per capita and overall meat consumption in the U.S. rose from 1970 to 2005, and then dropped slightly between 2005 and 2013. From 1970 to 2005 per capita consumption of beef and lamb, the types of meats linked to the greatest greenhouse gas emissions, dropped by about a third, and from

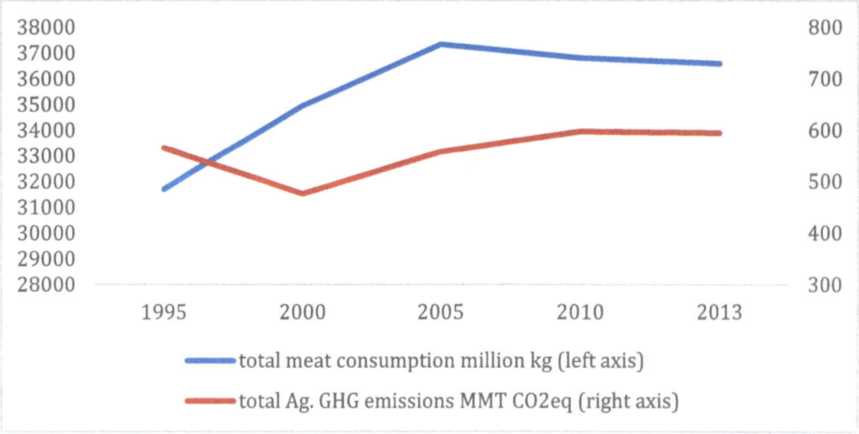

Fig. 3.3 U.S. meat consumption and agricultural greenhouse gas emissions 1995–2013. (Data from U.S. Department of Agriculture Economic Research Service, 2020)

[26] For example, (Garnett et al., 2017) state that "There are also other things we might want to do with the land currently used for grazing. A large portion of the grasslands livestock use has been formed from forest (they are not 'natural'), have little ecological value (consisting of fertilised grass monocultures) or may be degraded. That land could potentially be used for something else, including for nature conservation or for carbon sequestration by tree planting. At the same time, over a quarter of grassland is in fact suitable for cropping. There will be a carbon loss if the land is ploughed up and the negative impacts in the U.S. prairielands have already been noted. But since a relatively small area of cropland can produce the food output of a much larger area of grazing land, an argument could be made for sacrificing a small quantity of grassland to 'spare' the remaining grassland from intensification."

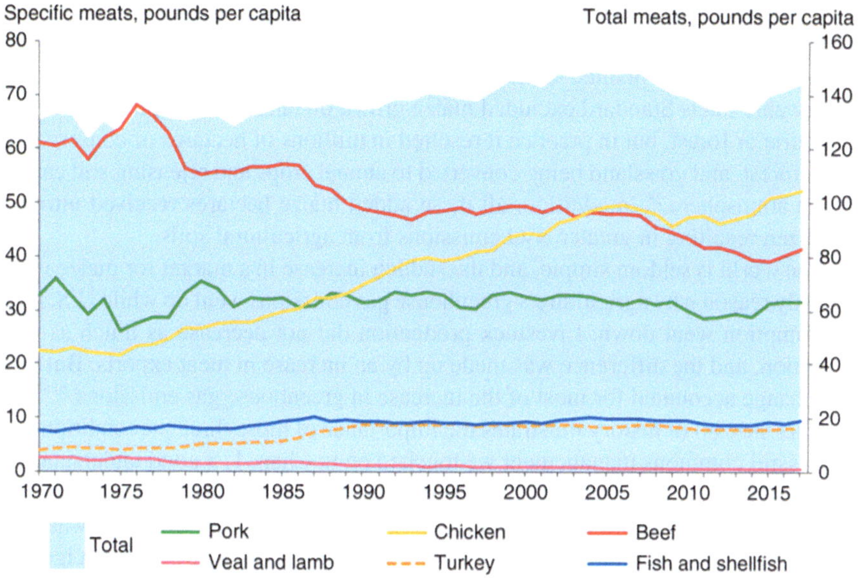

Source: USDA, Economic Research Service using loss-adjusted food availability data from its Food Availability (Per Capita) Data System.

Fig. 3.4 Per capita meat consumption trends in the U.S., 1970–2015. (Bentley 2019)

2005 to 2013 by another 14%. If more than half of agricultural greenhouse gas emissions in the U.S. are caused by livestock, and especially by beef, we would expect those emissions to rise when overall meat consumption was rising, and to decline when meat consumption went down. Yet from 1995 to 2000, as meat consumption was rising, agricultural greenhouse gas emissions dropped from 566 to 477 MMT. And from 2005 to 2013, as total meat consumption fell, U.S. agricultural greenhouse gas emissions rose from 558 to 595 MMT CO_2 eq.[27] How can this be?

One of the major factors driving emissions up was the emergence of a new market for maize that more than replaced the drop in demand for animal feed. American farmers have faced declining real prices for their products, and especially for the staple grain corn (maize), for many decades. During the 1980s and 90s federal programs paid farmers to put some farmland into conservation programs and plant native warm-season grasses or trees on that land. The goal of the conservation programs was twofold: to reduce erosion and water pollution, and to reduce overproduction of grain. The grasses and trees planted on conservation land helped store carbon in the soil, contributing to a slight decline in total greenhouse gas emissions.

Then in the early 2000s a spike in world petroleum prices, combined with pressure from American farm groups, resulted in the passage of the Renewable Fuels Standard in the Energy Policy Act of 2005. This act required gasoline to be blended

[27] Bentley (2019), U.S. Department of Agriculture Economic Research Service (2020).

with alcohol derived from plants, and the easiest and cheapest way to produce that alcohol was from maize. Suddenly the price for maize skyrocketed, and farmers across the country planted as much land as they could in corn. In theory, the Renewable Fuels Standard excluded maize grown on land that had previously been in prairie or forest, but in practice it resulted in millions of hectares of conservation land, forest, and grassland being converted to annual crops and releasing soil carbon to the atmosphere.[28] In addition, all these added maize hectares received nitrogen fertilizer, resulting in greater N_2O emissions from agricultural soils.

The world is seldom simple, and the sudden increase in a market for maize is not the only reason why agriculture's greenhouse gas emissions went up while U.S. meat consumption went down. Livestock production did not decrease as much as consumption, and the difference was made up by an increase in meat exports. But land use change accounted for most of the increase in greenhouse gas emissions.[29]

This little bit of history illustrates the importance of those three preconditions for successful commons management we touched on in Chap. 1: a good understanding of the ecological processes involved, honest negotiation, and fair treatment of all participants using the resource. In theory, the Renewable Fuel Standard was supposed to reduce the climate impact of transportation in the U.S. In practice it ignored the science that greenhouse gas emission reductions from substituting ethanol for petroleum were modest at best, and that the potential for land use change to cause major carbon emissions was significant. This disregard of science resulted from dishonest negotiation by key political players. And this dishonest negotiation was supported at least in part by desperate farmers whose livelihoods had been undermined by decades of policy that favored big farms over medium-sized ones, and agribusiness over farmers. The long decline in demand for beef and the grain to feed cattle almost certainly contributed to this desperation. Although chicken and pork emit less methane and CO_2 per kilo of meat than beef, most pork and poultry production does not offer a viable alternative for American livestock farmers. The growing market for chicken and pork is controlled from egg to supermarket by a handful of powerful meat processing companies, and farmers in that sector have fared even more poorly than those raising cattle and maize.[30]

This history is not meant to show that there is no climate benefit of reducing total livestock production. Rather, it shows that even when individual actions add up to significant change they need to be complemented by good policy to prevent unintended consequences. And good policy is much more likely to emerge if it is based on good science, honest negotiation, and fair treatment of all. This quote from an article in *Drovers*, a publication for cattle farmers and those who work with them, captures both the livestock industry's concern about climate policy and the potential for their participation in a process that they see as both realistic and fair:

[28] Lambert et al. (2020), Lark et al. (2015), Bigelow and Hellerstein (2020).

[29] U.S. Department of Agriculture (2016a, b) Table 1–2

[30] U.S. Department of Justice (2012), Hendrickson and James (2005).

> "My fear is that something like a cow tax will just (treat) all animals the same, no matter how they are raised and what part of the country they are raised in, no matter how they are fed, no matter how productive they are," he said. "I think that would be unproductive. We want to make sure if we go down that route that we have a system that actually incentivizes folks to move in a good direction."[31]

These examples are from the U.S., the place I know best. But I suspect the same is true of other parts of the world. If Brazil completely lost its beef export market would that stop or slow Amazonian deforestation? The answer would depend on a variety of factors, including the policy environment. If the underlying drivers for Amazonian deforestation are poverty and population growth, and if government policy does not prioritize forest protection, then a drop in global meat demand would not reduce deforestation, but would simply result in different, and possibly more damaging, land use in deforested areas, such as soybean or sugarcane production for liquid transportation fuel. Policy's influence on Amazonian deforestation can be seen in the sharp acceleration of deforestation that followed the election of Jair Bolsenaro as president of Brazil. Even before Bolsenaro took office the rate of illegal logging began to rise, because loggers rightly predicted that they would not be prosecuted under the new administration.[32]

3.4 The Need for Livestock in a Sustainable Agriculture

The university center where I work has the awkward name "Center for Integrated Agricultural Systems." In 1989, when the Center was founded (in response to pressure from frustrated farmers who demanded that the university pay attention to their needs and attempts to farm in ways that supported the environment and local communities), the words "sustainable" and "agroecology" were seen as too radical and divisive, and the idea of regenerative agriculture was not even discussed, at least in academic circles. So they wound up with "Integrated Agricultural Systems." People who pay attention to words almost always ask what "integrated" means in the name. The answer is multi-dimensional: agricultural systems are integrated in ecosystems and social systems, and also agricultural systems that integrate crops and livestock are more environmentally sustainable.

There are several ways that animal agriculture contributes to agricultural sustainability: it helps maintain soil fertility, it contributes to soil health and agricultural diversity, and it produces food on land that is not suited for growing crops.

---

[31] Henderson (2020).

[32] Escobar (2019), Lucas Ferrante and Fearnside (2019), Gibbs et al. (2016), Ferrante and Fearnside (2018).

3.4.1 Livestock and Soil Fertility

Plants use energy from sunlight, carbon from the air, and water from the soil to make carbohydrates, which humans and other animals then eat. But plants also need other chemicals to grow. Three of the most important plant nutrients are phosphorus, potassium, and nitrogen. Agricultural systems differ from wild ecosystems in that humans remove a lot of biomass from agricultural lands every year, and along with that biomass we remove the phosphorus, potassium, nitrogen, and other nutrients the plants took from the soil. So farmers have developed a variety of strategies to maintain the fertility of agricultural lands. The two main strategies farmers rely on, especially in Europe and North America, can be described as use of fertilizer mined or manufactured off the farm and organic/livestock-based fertility management.

Most farmers in Europe and North America buy mined phosphorus and potassium, and nitrogen pulled out of the air using the Haber-Bosch process to add to their soils.[33] These fertilizers are effective at boosting plant growth, and the yield gains of the Green Revolution are in part a result of a significant increase in production and use of these products in the second half of the twentieth century. But there are also costs of relying on mined phosphate, potassium, and especially synthetic (e.g., Haber-Bosch) nitrogen.

Phosphate (phosphorus) and potash (potassium) mines have a variety of negative impacts on local habitat, air quality, water quality, and human health.[34] Production of phosphate fertilizer is also energy-intensive and accounts for about 0.25% of the world's total energy use.[35] However, the energy needed to manufacture phosphates pales beside that needed to synthesize nitrogen fertilizer, which accounts for roughly 1.2% of the world's energy use and associated carbon emissions.[36] Moreover, fertilizer use is higher in Europe and North America than in less wealthy regions of the world, so it accounts for a greater portion of agriculture's greenhouse gas emissions. In Germany, for example, the emissions from producing nitrogen fertilizer may account for more than half of agriculture's total greenhouse gas emissions.[37]

Nitrogen fertilizer causes significant greenhouse gas emissions even after its manufacturing process. Usually about half of nitrogen that is applied to crops is taken up by the plants. The other half has a range of potential fates depending on temperature, soil moisture and other soil conditions: it can leach into groundwater (causing human health problems when it enters well-water), it can run off to surface

[33] Named for two German chemists, Haber-Bosch is the energy-intensive process of converting nitrogen in the air to ammonia, which can be used as nitrogen fertilizer. The production of nitrogen fertilizer through this process accounts for an estimated 1% of the world's fossil fuel use (Krietsch Boerner, 2019).

[34] Reta et al. (2018).

[35] Ramírez and Worrell (2006).

[36] Beste and Idel (2020).

[37] Beste and Idel (2020).

water (causing algae blooms and then low-oxygen conditions and fish kills in lakes and seas), or it can be used by soil microbes and then emitted to the air as ammonia or as nitrous oxide. Nitrous oxide or N_2O is an extremely powerful and long-lived greenhouse gas, and N_2O emissions from soils, largely due to nitrogen fertilizer, account for about 40% of agricultural greenhouse gas emissions in the U.S.[38]

For all these reasons organic agriculture prohibits the use of synthetic nitrogen fertilizer, as well as many forms of mined phosphates. Farmers who grow organic crops instead rely mostly on livestock manure, other livestock by-products such as feather meal and bone meal, and on green manures to maintain the fertility of their fields.

All manure contains significant amounts of phosphorus and potassium, and some nitrogen. The amount of nitrogen in manure depends on the type of livestock and how the manure is managed. One limitation of relying on manure as a fertilizer is that manure that has been stored usually contains more phosphorus and less nitrogen than the crop needs. In other words, if a farmer applies enough manure to a field to meet the crop's nitrogen needs, then that field will likely build up excessive levels of phosphorus. Agricultural phosphorus runoff is a major cause of water pollution and eutrophication of lakes.

This is where green manures come in. Green manure crops are plants that have a symbiotic relationship with soil bacteria that take nitrogen from the air and convert it to nitrates (NO_3), which is the form of nitrogen that plants can use. The crops that add the greatest amount of nitrogen to the soil – clovers, alfalfa and other legumes – are excellent feed for ruminants but are not digestible by humans. As a result, almost all organic farms rely on livestock by using manure and/or by feeding green manure crops or selling them for feed, even if they don't have any animals on the farm.[39] Green manures and other crops that only livestock can eat also reduce weed and pest populations that build up when the same crop or type of crop is grown year after year on the same soils.

It is important to note that the nitrogen from both animal manure and green manures can also be converted to nitrous oxides. Thus, it is critical to manage all fertilizers carefully, whether they are organic or synthetic.[40] Also, many farms use both manure and commercial/synthetic fertilizers. In the twentieth century the availability of cheap commercial fertilizer allowed crop farms to separate from livestock farms, but that separation came with some significant environmental costs. The best-known of these costs is the build-up of manure in large livestock farms that buy some or all of their feed. Because these farms don't grow all their own feed they don't have enough crop fields nearby that need the nutrients in the manure the farms produce. While it is cost-efficient to transport livestock feed in the form of grain over large distances, it is not cost-effective to transport manure very far. The quanti-

---

[38] Union of Concerned Scientists (2020), U.S. Department of Agriculture (2016a, b), U.S. Environmental Protection Agency (2020).

[39] Carr et al. (2019, 2020).

[40] VanderZaag et al. (2011), Dangal et al. (2019), Tian et al. (2020), Butterbach-Bahl et al. (2013).

ties of manure large livestock farms build up can damage the health of farm workers and of people living near the farms, and all too often pollute local lakes, rivers, and ground-water sources.[41] Of course, if we eliminated livestock production that would get rid of these problems of too much manure, but our food system would then also depend completely on mined and synthesized fertilizers.

This observation brings us back to the importance of integrating crops and livestock. If we do have livestock, we want to use their manure on the crops that feed both humans and animals. In other words, we want an almost closed nutrient cycle on our agricultural lands. When livestock and crop production are separated spatially, closing that nutrient cycle becomes more difficult.

3.4.2 Soil Health and Agricultural Diversity

Manure does more than just bring nutrients, though. When applied in appropriate ways and amounts it also supports a healthy soil biology and storage of soil carbon.[42] Increasing crop diversity by adding green manure crops also benefits soil health. And, unlike most grain and vegetable crops that humans can eat, pasture and range for ruminants usually has a diverse mix of perennial plants that keeps the soil covered year-round and also offers habitat for grassland wildlife. As a result, soil scientists and farmers often say that integrating livestock in the farming system is one of the foundational principles of soil health.

Two other principles of soil health are (1) keep the soil covered and (2) keep living roots in the soil.[43] The majority of U.S. cropland is in annual grains that are planted in in the spring and harvested in the fall, leaving almost no living plants in the field for about half the year. Government agencies and non-profit groups encourage farmers to plant winter cover crops so that the soil surface will be covered by plants and held by living roots throughout the year. There are a few cold-hardy crops, like cereal rye, that can survive even bitter cold winter weather. Farmers want to protect their soil, but for many there is a hitch. If they plant the highest yielding variety of grain and wait until after harvest to plant a winter cover crop, then it is usually too late in the year for that winter cover to grow much. If the winter cover crop doesn't grow much, then it also doesn't protect the soil, so the farmer has the expense and labor of planting a cover crop but doesn't get the expected environmen-

[41] Greger and Koneswaran (2010), Burkholder et al. (2007), Carrel et al. (2016), Ashwood et al. (2014).

[42] Rayne and Aula (2020).

[43] Conservationists, agricultural educators, and farmers typically list either four or five principles of soil health. There is agreement on the benefits of limiting soil disturbance, keeping the soil covered, keeping living roots in the ground, and having a diverse plant community, but while some stress the importance of integrating animals in the farming system to support soil health, others do not include that fifth principle (South Dakota Soil Health Coalition, n.d.; U.S. Department of Agriculture n.d.).

tal benefit. If the farmer plants a grain that matures more quickly, that would allow an earlier harvest and time to establish a good winter crop. The problem is that in general early maturing varieties have lower yields, and of course most farmers are reluctant to reduce their income due to lower yields. Similarly, some farmers are experimenting with planting cover crops while the main cash crop is still growing, which is expected to improve cover crop performance in exchange for a small reduction in yield. But farmers who have livestock can get economic value from their winter cover crops by grazing them in early spring. The value of this feed for their livestock makes it economically viable for them to slightly reduce the yield from their summer cash crops to accommodate successful winter cover crops.[44]

Livestock also help maintain traditional habitats and manage the interface of human and wild landscapes. In the Mediterranean managed grazing of forests has historically been a tool to reduce the risk of forest fires.[45] As forest fires rage in the western U.S., releasing large amounts of carbon, endangering human health, and destroying property, there is increasing attention to reducing fire risk near human settlements. Redesigning the human-forest interface in these fire-prone areas is a complex challenge, but one part of the solution might be to have buffers of forested range between human settlements and wild forest, where grazing livestock help keep down fuel loads, while scattered trees provide aesthetic value and wildlife habitat and also store carbon.

In my home state of Wisconsin before European settlement the southwestern third of the state was a mix of savanna, tall-grass prairie, and some areas of forest.[46] The forest was mainly limited to river valleys and north-facing slopes. The Potawatomi and Ho-Chunk peoples that lived in this area managed the landscape to attract and sustain a mix of grassland and forest ruminants. Today forest, maize, soybeans, and alfalfa dominate the region, and only about 1 or 2% of the earlier savanna habitat remains, mostly on grazing farms. Grazing farms also provide the main habitat for meadowlarks, bobolinks, and other grassland birds, and both landowners and the state Department of Natural Resources are experimenting with using grazing as a tool to restore and maintain both savanna and prairie habitat.[47]

3.5 Marginal Land

Another important niche that livestock occupies is the ability to use land that is not suitable for crop production because it is too steep or too rocky or too dry or cold. Think of Swiss cows and goats grazing alpine pastures, migratory cattle herds ranging across East Africa, sheep grazing England's downs, Mongolian camels, and

[44] Myers et al. (2019).

[45] Rubino (1996), Bland and Auclair (1996).

[46] Cottam and Loucks (1965).

[47] Harrington and Kathol (2009), Anderson (2020).

reindeer in Sami lands. But even in areas that have great cropland many farms contain some land that is not suited for growing vegetables or grains. Some of those marginal lands should be left as natural areas, but in others livestock can provide income and maintain environmental quality. For example, Harmony Valley Farm is an organic vegetable farm in southwestern Wisconsin. The farm is located in the valleys of a creek and small river, with flat, fertile floodplain soils next to the water, flanked by steep hills. The farm grows an amazing array of vegetables in the valley bottom and leaves the steep slopes in forest. But on the gently sloping transition between the vegetable fields and the forest they graze steers. These steers keep the lower slopes open so the vegetable fields get enough sun, and at the same time they provide the farm with a little supplemental income.

3.6 Cultural Value of Meat

When I was a recent college graduate in 1977 I volunteered for an environmental organization that was making a major push to reduce the hunting of marine mammals. That seemed like a very worthy cause to me, but I was shocked to find that this environmental group advocated allowing Inuit communities to continue to hunt whales in their proposed anti-whale hunting legislation. Bowhead whales were highly endangered – why would a law designed to protect them explicitly allow a certain group of people to kill them? As I slowly came to realize, the answer is based in two issues of fairness. First, it is not Inuit hunting that brought about the worldwide decline in marine mammal populations – that was caused by a combination of industrial whaling, ocean pollution, modern fishing methods that unintentionally swept up and killed marine mammals, heavy ocean traffic that disrupted nurseries and migration routes, and other side effects of industrial economies. And second was the huge importance of whale hunting, sharing, and eating in Inuit culture. In a community already under huge stress it was not right to take away an act with such critical spiritual value and importance for community cohesion and healthy nutrition.

Similarly, cattle play important cultural and community roles for many people around the world. Consider AmaXhosa villagers in South Africa's Eastern Cape district. Unlike residents of wealthy nations, they don't take plane trips or own cars or heat, cool, and power 185 m^2 (2000 square foot) houses for a household of 2.3 people. So the cows they own may account for half or more of their climate footprint, especially since those cows subsist on poor quality forage, which can result in higher methane emissions than a diet of high quality forage or a diet with high carbohydrate supplements like grain or molasses.[48] But just as Inuit hunting did not cause the crash in marine mammal populations, so these villagers each have a much smaller carbon footprint than almost all of us who live in the U.S. or Canada or western Europe. And their cattle have high cultural importance. Their cows connect

[48] van Gastelen et al. (2019).

them to their ancestors, whom they honor by sacrificing animals at important life ceremonies like the birth and coming of age of a child.[49] These meat-based feasts bring families and communities together and provide important nutrients in diets dominated by cornmeal porridge. On a more pedestrian note, people in this community have little access to banks, and cattle are how they have traditionally stored their wealth. Perhaps as cell phone service improves they may have other options for building wealth, but for now many of them store their savings as cows and sheep.

Or let's look at the case of India. Meat consumption in India is extremely low, only about 5 kg per year per person, and most of that meat is poultry. Yet methane from ruminants accounts for about half of India's total greenhouse gas emissions.[50] Indians do eat dairy products, making the country the world's leading consumer of milk. While per capita dairy consumption is below that of the U.S. and northern Europe, it is growing and is projected to equal European levels in coming decades.[51] Dairy production is also an important source of income for about 80 million small farmers. And of course, cattle have great cultural and spiritual significance, especially for the majority Hindu religion. I would argue that India needs to figure out how the country might reduce ruminant methane emissions in ways that honor the culture and spiritual needs of their diverse population, support the health of the people, and maintain or improve the livelihoods of smallholders. Ending meat and dairy production and consumption won't meet those goals, and so is both politically unlikely and ethically questionable.

3.7 Livestock and Food Security

We need food to live, but food production is uncertain. When the weather cooperates farmers get high yields, and when the weather doesn't cooperate they get low yields. This unpredictability is one reason why agricultural policy errs on the side of promoting excess production. Bad as the consequences of overproduction are, they are preferable to the impacts of not having enough food, at least in the short term. And food takes time to grow. Just-in-time production may work for many manufactured goods, but if we suddenly find we need more food, for most crops we will have to wait a year to get to the next harvest. One of the important roles of livestock throughout history has been to use the excess agricultural production in good years, and to be expendable in bad years. With climate change we are seeing more variation in weather patterns challenging agriculture, from years of drought in some parts of the world, to repeated flooding and damaging storms in others. In addition to weather, other factors like changes in pest and disease pressures add even more

[49] Bell (2018). Villagers trace the lineages of all their cattle to the ancestors who contributed them.

[50] Ganesan et al. (2017), Devi et al. (2014), OECD/FAO (2017).

[51] Aradhey (2020).

uncertainty. Even with modern agricultural tools we still need to plan and plant for lower than expected yields, and let livestock use the excess when yields are good.[52]

3.8 Diet and Health

A lot has been written about the role of livestock products in a healthy and sustainable diet. Not surprisingly, what the studies conclude depends in large part on their assumptions about the details of the diets they are comparing. Here are a few observations that can help frame our thinking around the intersection of health, food, and climate change: Many people do well on a vegan diet, but dietary needs and limitations vary among individuals and groups, so the fact that some people thrive on a vegan diet does not prove that it is healthy for all. The relationship between diet healthfulness and greenhouse gas emissions is not linear. A number of studies have concluded that diets with the lowest greenhouse gas emissions are not healthy, because they rely heavily on grains and do not include healthy amounts of fruits and vegetables, which tend to have much higher emissions than grains. At the same time, there is little question that in general diets that are very high in livestock products have higher greenhouse gas emissions and typically do not improve health outcomes compared to diets with moderate quantities of livestock products and more moderate emissions.[53] Finally, food is about more than nutrients and calories – it also has great cultural importance, and we are increasingly coming to understand that food culture also has implications for health.

3.9 Concluding Thoughts

Thinking back to my neighbor's question on that snowy morning, it isn't that easy to know what meat is best to eat from a climate perspective. Poultry typically has much lower greenhouse gas emissions than beef, but with the right management ruminants can support soil health and sustainable cropping systems and can make better use of marginal land than chickens.

Certainly, there is room for North American and European nations to reduce per capita meat consumption, and such a reduction, if accompanied by good policy, could reduce agriculture's greenhouse gas emissions. Thus, a thoughtful reduction in the amount of livestock raised and the amount of meat consumed can provide a helpful contribution to mitigating climate change, though it does not relieve the

[52] While that expendable role of livestock is good for society, it can be hard on livestock farmers, especially farms that are not diversified and specialize in raising animals (Isai & Gundlock, 2021).

[53] Tom et al. (2016), Auestad and Fulgoni (2015).

pressing need to reduce fossil fuel production and consumption for transportation, electricity generation, cement production, and heating and cooling.

However, eliminating livestock agriculture altogether is not a culturally or environmentally beneficial goal, as well as not being realistic. This makes it critical to understand how we can improve livestock production methods to sustainably reduce associated greenhouse gas emissions.

Bibliography

Alig, R. J., Plantinga, A. J., Haim, D., & Todd, M. (2010). *Area Changes in U.S. Forests and Other Major Land Uses, 1982 to 2002, With Projections to 2062* (PNW-GTR-815). US Department of Agriculture Forest Service.

Anderson, M. C. (2020). *Grazing public lands in Wisconsin*. Wisconsin Department of Natural Resources. https://pastureproject.org/wp-content/uploads/2020/07/Grazing-Public-Lands.pdf

Aradhey, A. (2020). *India: Dairy and products annual*. USDA Foreign Agricultural Service. https://www.fas.usda.gov/data/india-dairy-and-products-annual-4

Asher, K., Green, C., Gutbrod, Mirna Jewell, Dr Galina Hale Dr Brock Bastian, H., Jewell, M., Hale, G., Bastian, B. (2014). *Study of current and former vegetarians and vegans* (p. 10).

Ashwood, L., Diamond, D., & Thu, K. (2014). Where's the farmer? Limiting liability in midwestern industrial hog production. *Rural Sociology, 79*(1), 2–27. https://doi.org/10.1111/ruso.12026

Auestad, N., & Fulgoni, V. L. (2015). What current literature tells us about sustainable diets: Emerging research linking dietary patterns, environmental sustainability, and economics. *Advances in Nutrition, 6*(1), 19–36. https://doi.org/10.3945/an.114.005694

Bell, M. M. (2018). *City of the good: Nature, religion, and the ancient search for what is right*. Princeton University Press.

Bentley, J. (2017). *U.S. Trends in Food availability and a dietary assessment of loss-adjusted food availability, 1970–2014* (EIB-166). U.S. Department of Agriculture, Economic Research Service.

Bentley, J. (2019, December 2). U.S. per capita availability of red meat, poultry, and seafood on the rise. *Amber waves*. https://www.ers.usda.gov/amber-waves/2019/december/us-per-capita-availability-of-red-meat-poultry-and-seafood-on-the-rise/

Beste, A., & Idel, A. (2020, June 1). The myth of climate smart agriculture – Why Less Bad Isn't Good. *ARC* 2020. https://www.arc2020.eu/the-myth-of-climate-smart-agriculture-why-less-bad-isnt-good/

Bigelow, D., & Hellerstein, D. (2020, February 3). In recent years, most expiring land in the conservation reserve program returned to crop production. *Amber Waves*. https://www.ers.usda.gov/amber-waves/2020/february/in-recent-years-most-expiring-land-in-the-conservation-reserve-program-returned-to-crop-production/

Bland, F., & Auclair, D. (1996). Silvopastoral aspects of Mediterranean forest management. In *Western European silvopastoral systems* (pp. 125–141). INRA (Institut National de la Recherche Agronomique).

Bohrer, B. M. (2017). Review: Nutrient density and nutritional value of meat products and non-meat foods high in protein. *Trends in Food Science & Technology, 65*, 103–112. https://doi.org/10.1016/j.tifs.2017.04.016

Bureau of Transportation Statistics. (n.d.). *Table 4-23: Average fuel efficiency of U.S. light duty vehicles*. U.S. Department of Transportation. https://www.bts.gov/content/average-fuel-efficiency-us-light-duty-vehicles

Burkholder, J., Libra, B., Weyer, P., Heathcote, S., Kolpin, D., Thorne, P. S., & Wichman, M. (2007). Impacts of waste from concentrated animal feeding operations on water quality. *Environmental Health Perspectives, 115*(2), 308–312. https://doi.org/10.1289/ehp.8839

Butterbach-Bahl, K., Baggs, E. M., Dannenmann, M., Kiese, R., & Zechmeister-Boltenstern, S. (2013). Nitrous oxide emissions from soils: How well do we understand the processes and their controls? *Philosophical Transactions of the Royal Society B: Biological Sciences, 368*(1621), –20130122. https://doi.org/10.1098/rstb.2013.0122

Carr, P. M., Cavigelli, M. A., Darby, H., Delate, K., Eberly, J. O., Gramig, G. G., Heckman, J. R., Mallory, E. B., Reeve, J. R., Silva, E. M., Suchoff, D. H., & Woodley, A. L. (2019). Nutrient cycling in organic field crops in Canada and the United States. *Agronomy Journal, 111*(6), 2769–2785. https://doi.org/10.2134/agronj2019.04.0275

Carr, P. M., Cavigelli, M. A., Darby, H., Delate, K., Eberly, J. O., Fryer, H. K., Gramig, G. G., Heckman, J. R., Mallory, E. B., Reeve, J. R., Silva, E. M., Suchoff, D. H., & Woodley, A. L. (2020). Green and animal manure use in organic field crop systems. *Agronomy Journal, 112*(2), 648–674. https://doi.org/10.1002/agj2.20082

Carrel, M., Young, S., & Tate, E. (2016). Pigs in space: Determining the environmental justice landscape of swine concentrated animal feeding operations (CAFOs) in Iowa. *International Journal of Environmental Research and Public Health, 13*(9), 849. https://doi.org/10.3390/ijerph13090849

Carrington, D. (2018, May 31). Avoiding meat and dairy is 'single biggest way' to reduce your impact on earth. *The Guardian*. https://www.theguardian.com/environment/2018/may/31/avoiding-meat-and-dairy-is-single-biggest-way-to-reduce-your-impact-on-earth

Clune, S., Crossin, E., & Verghese, K. (2017). Systematic review of greenhouse gas emissions for different fresh food categories. *Journal of Cleaner Production, 140*, 766–783. https://doi.org/10.1016/j.jclepro.2016.04.082

Cottam, G., & Loucks, O. L. (1965). *Early vegetation of Wisconsin*. Wisconsin Geological and Natural History Survey. https://wgnhs.wisc.edu/catalog/publication/000360.

Dangal, S. R. S., Tian, H., Xu, R., Chang, J., Canadell, J. G., Ciais, P., Pan, S., Yang, J., & Zhang, B. (2019). Global nitrous oxide emissions from pasturelands and rangelands: Magnitude, spatiotemporal patterns, and attribution. *Global Biogeochemical Cycles, 33*(2), 200–222. https://doi.org/10.1029/2018GB006091

Devi, S. M., Balachandar, V., Lee, S. I., & Kim, I. H. (2014). An outline of meat consumption in the Indian population—A pilot review. *Korean Journal for Food Science of Animal Resources, 34*(4), 507–515. https://doi.org/10.5851/kosfa.2014.34.4.507

Doll, J. E., Petersen, B., & Bode, C. (2017). Skeptical but adapting: What midwestern farmers say about climate change. *Weather, Climate, and Society, 9*(4), 739–751. https://doi.org/10.1175/WCAS-D-16-0110.1

Escobar, H. (2019). Brazil's deforestation is exploding—And 2020 will be worse. *Science, 22*(Nov). https://doi.org/10.1126/science.aba3238

Eshel, G., Shepon, A., Makov, T., & Milo, R. (2014). Land, irrigation water, greenhouse gas, and reactive nitrogen burdens of meat, eggs, and dairy production in the United States. *Proceedings of the National Academy of Sciences, 111*(33), 11996–12001. https://doi.org/10.1073/pnas.1402183111

Eshel, G., Shepon, A., Shaket, T., Cotler, B. D., Gilutz, S., Giddings, D., Raymo, M. E., & Milo, R. (2018). A model for 'sustainable' US beef production. *Nature Ecology & Evolution, 2*(1), 81–85. https://doi.org/10.1038/s41559-017-0390-5

Ferrante, L., & Fearnside, P. M. (2018). Amazon sugar cane: A threat to the forest. *Science, 359*(6383), 1476–1476. https://doi.org/10.1126/science.aat4208

Ferrante, L., & Fearnside, P. M. (2019). Brazil's new president and 'ruralists' threaten Amazonia's environment, traditional peoples and the global climate. *Environmental Conservation, 46*(4), 261–263. https://doi.org/10.1017/S0376892919000213

Ganesan, A. L., Rigby, M., Lunt, M. F., Parker, R. J., Boesch, H., Goulding, N., Umezawa, T., Zahn, A., Chatterjee, A., Prinn, R. G., Tiwari, Y. K., van der Schoot, M., & Krummel, P. B. (2017).

Atmospheric observations show accurate reporting and little growth in India's methane emissions. *Nature Communications, 8*(1), 836. https://doi.org/10.1038/s41467-017-00994-7

Garnett, T., Godde, C., Muller, A., Röös, E., Smith, P., & de Boer, I. (2017). *Ruminating on cattle, grazing systems, methane, nitrous oxide, the soil carbon sequestration question – and what it all means for greenhouse gas emissions*, p. 127.

Gerber, P. J., Hristov, A. N., Henderson, B., Makkar, H., Oh, J., Lee, C., Meinen, R., Montes, F., Ott, T., Firkins, J., Rotz, A., Dell, C., Adesogan, A. T., Yang, W. Z., Tricarico, J. M., Kebreab, E., Waghorn, G., Dijkstra, J., & Oosting, S. (2013). Technical options for the mitigation of direct methane and nitrous oxide emissions from livestock: A review. *Animal, 7*(s2), 220–234. https://doi.org/10.1017/S1751731113000876

Gibbs, H. K., Munger, J., L'Roe, J., Barreto, P., Pereira, R., Christie, M., Amaral, T., & Walker, N. F. (2016). Did ranchers and slaughterhouses respond to zero-deforestation agreements in the Brazilian Amazon?: Brazil's zero-deforestation pacts. *Conservation Letters, 9*(1), 32–42. https://doi.org/10.1111/conl.12175

Greger, M., & Koneswaran, G. (2010). The public health impacts of concentrated animal feeding operations on local communities. *Family & Community Health, 33*(1), 11–20. https://doi.org/10.1097/FCH.0b013e3181c4e22a

Harrington, J. A., & Kathol, E. (2009). Responses of shrub midstory and herbaceous layers to managed grazing and fire in a North American Savanna (Oak woodland) and prairie landscape. *Restoration Ecology, 17*(2), 234–244. https://doi.org/10.1111/j.1526-100X.2008.00369.x

Harwatt, H., Sabaté, J., Eshel, G., Soret, S., & Ripple, W. (2017). Substituting beans for beef as a contribution toward US climate change targets. *Climatic Change, 143*(1–2), 261–270. https://doi.org/10.1007/s10584-017-1969-1

Haverstock, K., & Forgays, D. K. (2012). To eat or not to eat. A comparison of current and former animal product limiters. *Appetite, 58*(3), 1030–1036. https://doi.org/10.1016/j.appet.2012.02.048

Helander, H., Bruckner, M., Leipold, S., Petit-Boix, A., & Bringezu, S. (2021). Eating healthy or wasting less? Reducing resource footprints of food consumption. *Environmental Research Letters, 16*(5), 054033. https://doi.org/10.1088/1748-9326/abe673

Henderson, G. (2020, October 12). Might the election affect meat demand? *Drovers*. https://www.drovers.com/article/might-election-affect-meat-demand

Hendrickson, M. K., & James, H. S. (2005). The ethics of constrained choice: How the industrialization of agriculture impacts farming and farmer behavior. *Journal of Agricultural and Environmental Ethics, 18*(3), 269–291. https://doi.org/10.1007/s10806-005-0631-5

Hoekstra, A. Y., & Mekonnen, M. M. (2012). The water footprint of humanity. *Proceedings of the National Academy of Sciences, 109*(9), 3232–3237. https://doi.org/10.1073/pnas.1109936109

Hoekstra, A. Y., Mekonnen, M. M., Chapagain, A. K., Mathews, R. E., & Richter, B. D. (2012). Global monthly water scarcity: Blue water footprints versus blue water availability. *PLoS One, 7*(2), e32688. https://doi.org/10.1371/journal.pone.0032688

Hu, K., & Chen, Y. (2016). Technological growth of fuel efficiency in European automobile market 1975–2015. *Energy Policy, 98*, 142–148. https://doi.org/10.1016/j.enpol.2016.08.024

Isai, V., & Gundlock, B. (2021, August 4). Farmers race to save their cattle from Canada's drought. *New York Times*. https://www.nytimes.com/2021/08/04/world/canada/manitoba-drought-cattle-farms.html

Katz-Rosene, R. M. (2020). How do livestock impact the climate. In *Green meat?: Sustaining eaters animals and the planet* (pp. 43–63). McGill-Queen's University Press.

Krietsch Boerner, L. (2019, June 15). Industrial ammonia production emits more CO2 than any other chemical-making reaction. Chemists want to change that. *Chemical & Engineering News, 97*(24). https://cen.acs.org/environment/green-chemistry/Industrial-ammonia-production-emits-CO2/97/i24

Lambert, A., Hallar, A. G., Garcia, M., Strong, C., Andrews, E., & Hand, J. L. (2020). Dust impacts of rapid agricultural expansion on the Great Plains. *Geophysical Research Letters, 47*(20). https://doi.org/10.1029/2020GL090347

Lark, T. J., Salmon, J. M., & Gibbs, H. K. (2015). Cropland expansion outpaces agricultural and biofuel policies in the United States. *Environmental Research Letters, 10*(4), 044003. https://doi.org/10.1088/1748-9326/10/4/044003

Mayerfeld, D., Rickenbach, M., & Rissman, A. (2016). Overcoming history: Attitudes of resource professionals and farmers toward silvopasture in Southwest Wisconsin. *Agroforestry Systems, 90*(5), 723–736. https://doi.org/10.1007/s10457-016-9954-7

Myers, R., Weber, A., & Tellatin, S. (2019). *Cover crop economics: Opportunities to improve your bottom line in row crops.* U.S. Department of Agriculture, Sustainable Agriculture Research and Education. https://www.sare.org/wp-content/uploads/Cover-Crop-Economics.pdf.

Nepstad, D., Soares-Filho, B. S., Merry, F., Lima, A., Moutinho, P., Carter, J., Bowman, M., Cattaneo, A., Rodrigues, H., Schwartzman, S., McGrath, D. G., Stickler, C. M., Lubowski, R., Piris-Cabezas, P., Rivero, S., Alencar, A., Almeida, O., & Stella, O. (2009). The end of deforestation in the Brazilian Amazon. *Science, 326*(5958), 1350–1351. https://doi.org/10.1126/science.1182108

OECD. (2021). *Meat consumption* [OECD Data]. https://data.oecd.org/agroutput/meat-consumption.htm

OECD/FAO. (2017). *Agricultural Outlook 2017-2026*. OECD Publishing. https://doi.org/10.1787/agr_outlook-2017-en

Oswalt, S. N., & Smith, W. B. (2014). *U.S. Forest resource facts and historical trends* (FS-1035). US Department of Agriculture Forest Service. https://mff.forest.mtu.edu/TreeBasics/USFSfacts&trends2014.pdf

Pimentel, D., & Pimentel, M. (2003). Sustainability of meat-based and plant-based diets and the environment. *The American Journal of Clinical Nutrition, 78*(3), 660S–663S. https://doi.org/10.1093/ajcn/78.3.660S

Pimentel, D., Houser, J., Preiss, E., White, O., Fang, H., Mesnick, L., Barsky, T., Tariche, S., Schreck, J., & Alpert, S. (1997). Water resources: Agriculture, the environment, and society. *Bioscience, 47*(2), 97–106. https://doi.org/10.2307/1313020

Ramírez, C. A., & Worrell, E. (2006). Feeding fossil fuels to the soil. *Resources, Conservation and Recycling, 46*(1), 75–93. https://doi.org/10.1016/j.resconrec.2005.06.004

Rayne, N., & Aula, L. (2020). Livestock manure and the impacts on soil health: A review. *Soil Systems, 4*(4), 64. https://doi.org/10.3390/soilsystems4040064

Reta, G., Dong, X., Li, Z., Su, B., Hu, X., Bo, H., Yu, D., Wan, H., Liu, J., Li, Y., Xu, G., Wang, K., & Xu, S. (2018). Environmental impact of phosphate mining and beneficiation: Review. *International Journal of Hydrology, 2*(4), 10.15406/ijh.2018.02.00106.

Ridoutt, B. G., Sanguansri, P., Freer, M., & Harper, G. S. (2012). Water footprint of livestock: Comparison of six geographically defined beef production systems. *The International Journal of Life Cycle Assessment, 17*(2), 165–175. https://doi.org/10.1007/s11367-011-0346-y

Rose, D., Heller, M., & Willits-Smith, A. (2020). The water scarcity footprint of the US diet. *Current Developments in Nutrition, 4*(Suppl_2), 145–145. https://doi.org/10.1093/cdn/nzaa042_010

Rotz, C. A., Asem-Hiablie, S., Place, S., & Thoma, G. (2019). Environmental footprints of beef cattle production in the United States. *Agricultural Systems, 169*, 1–13. https://doi.org/10.1016/j.agsy.2018.11.005

Rubino, R. (1996). Forest grazing: Reflections on its evolution and the future. In M. Étienne (Ed.), *Western European silvopastoral systems* (pp. 157–165). INRA (Institut National de la Recherche Agronomique).

South Dakota Soil Health Coalition. (n.d.). *The five principles of soil health.* South Dakota Soil Health Coalition. https://www.sdsoilhealthcoalition.org/technical-resources/the-five-principles-of-soil-health/

Steinfeld, H., Gerber, P., Wassenaar, T. D., Castel, V., Rosales, M. M., & de Haan, C. (2006). *Livestock's long shadow: Environmental issues and options*. Food and Agriculture Organization of the United Nations.

Sweers, R. (2013). *Dried beans, peas, and lentils can help you save $*. Iowa State University. https://www.extension.iastate.edu/dallas/sites/www.extension.iastate.edu/files/dallas/N3515.pdf

Tian, H., Xu, R., Canadell, J. G., Thompson, R. L., Winiwarter, W., Suntharalingam, P., Davidson, E. A., Ciais, P., Jackson, R. B., Janssens-Maenhout, G., Prather, M. J., Regnier, P., Pan, N., Pan, S., Peters, G. P., Shi, H., Tubiello, F. N., Zaehle, S., Zhou, F., … Yao, Y. (2020). A comprehensive quantification of global nitrous oxide sources and sinks. *Nature, 586*(7828), 248–256. https://doi.org/10.1038/s41586-020-2780-0

Tietge, U., Mock, P., Franco, V., & Zacharof, N. (2017). From laboratory to road: Modeling the divergence between official and real-world fuel consumption and CO2 emission values in the German passenger car market for the years 2001–2014. *Energy Policy, 103*, 212–222. https://doi.org/10.1016/j.enpol.2017.01.021

Tom, M. S., Fischbeck, P. S., & Hendrickson, C. T. (2016). Energy use, blue water footprint, and greenhouse gas emissions for current food consumption patterns and dietary recommendations in the US. *Environment Systems and Decisions, 36*(1), 92–103. https://doi.org/10.1007/s10669-015-9577-y

U.S. Department of Agriculture. (2016a). *U. S. Agriculture and Forestry Greenhouse Gas Inventory, 1990–2013* (Technical Bulletin No. 1943). US Department of Agriculture. https://doi.org/10.15482/USDA.ADC/1264344

U.S. Department of Agriculture. (2016b). *U. S. Agriculture and Forestry Greenhouse Gas Inventory, 1990–2013*. U.S. Department of Agriculture, Office of the Chief Economist, Global Change Program Office. https://www.usda.gov/sites/default/files/documents/USDA_GHG_Inventory_1990-2013_9_19_16_reduced.pdf

U.S. Department of Agriculture Economic Research Service. (2020). *Meat supply and disappearance: Meat production, imports, total supply, exports, total and per capita disappearance with history*. https://www.ers.usda.gov/data-products/livestock-meat-domestic-data/livestock-meat-domestic-data/#All%20supply%20and%20disappearance

U.S. Department of Agriculture NRCS. (2013). *2007 National Resources Inventory: Land use status and trends*. https://www.nrcs.usda.gov/Internet/FSE_DOCUMENTS/stelprdb1083428.pdf

U.S. Department of Energy and U.S. Environmental Protection Agency. (n.d.). *The official U.S. government source for fuel economy information*. www.Fueleconomy.Gov. https://www.fueleconomy.gov/feg/PowerSearch.do?action=noform&path=1&year1=2020&year2=2020&make=Toyota&baseModel=Prius&srchtyp=ymm

U.S. Department of Justice. (2012). Competition and agriculture: Voices from the workshops on agriculture and antitrust enforcement in our 21st century economy and thoughts on the way forward. U.S. Department of Justice. https://www.justice.gov/sites/default/files/atr/legacy/2012/05/16/283291.pdf

U.S. Environmental Protection Agency. (2020). *Inventory of U.S. Greenhouse Gas Emissions and Sinks 1990–2018*. https://www.epa.gov/ghgemissions/sources-greenhouse-gas-emissions

Union of Concerned Scientists. (2020). Each country's share of CO2 emissions. https://www.ucsusa.org/resources/each-countrys-share-co2-emissions

van Gastelen, S., Dijkstra, J., & Bannink, A. (2019). Are dietary strategies to mitigate enteric methane emission equally effective across dairy cattle, beef cattle, and sheep? *Journal of Dairy Science, 102*(7), 6109–6130. https://doi.org/10.3168/jds.2018-15785

VanderZaag, A. C., Jayasundara, S., & Wagner-Riddle, C. (2011). Strategies to mitigate nitrous oxide emissions from land applied manure. *Animal Feed Science and Technology, 166–167*, 464–479. https://doi.org/10.1016/j.anifeedsci.2011.04.034

Chapter 4
The Benefits of Modern Efficiency

Diane Mayerfeld and Jude L. Capper

Abstract The greenhouse gas emissions produced by livestock vary widely, depending on species and production practices employed. Scientific advances in husbandry and on-farm practices, including artificial insemination, vaccination, housing, balanced nutrition and mineral supplements have greatly reduced the greenhouse gas emission intensity of beef and dairy in Europe and North America, both compared to the past and to other parts of the world. For example, milk production per cow has more than doubled in the U.S. since the middle of the twentieth century, while the growth rates and meat yields of beef cattle have also increased substantially. As a result of these productivity gains, on average, a kilo of beef or milk produced in Europe, North America, or Oceania now results in less than half the greenhouse gas emissions of a kilo of milk or beef from Latin America, Sub-Saharan Africa, or South Asia.

Keywords Climate change · Emissions intensity · Feed efficiency · Greenhouse gas emissions · Livestock

4.1 Introduction

As outlined in Chap. 3, there are important reasons to include livestock in our agricultural systems, and meat and dairy in our diets. However, the question then becomes: what is the best way to raise livestock? People often assume that traditional, entirely pasture-based livestock must be the most environmentally benign approach, but a strong body of research documents that, done with care, modern

---

D. Mayerfeld (✉)
Division of Extension, Agriculture Institute, University of Wisconsin, Madison, WI, USA
e-mail: dbmayerfeld@wisc.edu

J. L. Capper
Agriculture and Environment, Faculty Member, Harper Adams University, Shropshire, UK

D. Mayerfeld (ed.), *Our Carbon Hoofprint*, Food and Health,
https://doi.org/10.1007/978-3-031-09023-3_4

intensive livestock production practices can reduce the climate impacts of meat and dairy.

Let's take a moment to review some critical terminology. First, there are several possible ways one can quantify the greenhouse gas emissions of livestock. The most commonly used measure, and the one this chapter uses, is **emissions intensity** expressed as kg of carbon dioxide equivalents (CO_2eq) per kilogram of product (milk or meat). This measure is the sum of the global warming potentials (converted to kilograms of carbon dioxide equivalents, CO_2eq) of the carbon dioxide (CO_2), methane (CH_4), and nitrous oxide (N_2O) emitted in producing a kilogram of the product.[1] These greenhouse gases vary in their potential to cause global warming, with CO_2 used as the reference gas, with a global warming potential averaged over 100 years (GWP_{100}) set at a value of 1.0. Methane (CH_4) is 28 times more potent ($GWP_{100} = 28$) and nitrous oxide (N_2O) is 298 times more potent ($GWP_{100} = 298$) than CO_2.[2] Using their GWP values allows us to measure the total impact and compare to other products.

The GWP_{100} has been challenged by researchers from Oxford University who argued that, unlike GWP_{100}, which measures the accumulated radiative forcing effects of CO_2, CH_4 and N_2O over 100 years, a new metric called GWP* better accounts for the climatic temperature response to short-lived greenhouse gasses, such as CH_4.[3] If this metric was globally adopted, the relative GWP of CH_4 would decrease by approximately 75%, with a significant reduction in the total C footprint from livestock systems. In essence, livestock systems that have been in place for 20+ years could conceivably become carbon-neutral, rather than being net greenhouse gas emitters.[4] At the time of writing, however, GWP* has not been widely adopted by international climate change assessment bodies. Regardless of whether GWP_{100} or GWP* is used, using the kg CO_2eq metric connects the two things we are most concerned about: the foods that we humans want, and the greenhouse gases that we are releasing in order to produce those foods.

A mass-based denominator (kg, ton) is often used, yet we can also use metrics that account for the nutritional value of those products, e.g. protein, amino acids, or all nutrients expressed as an index.[5] For example, focusing on protein can correct for the fact that different animal products have different concentrations of nutrients; meat typically contains almost eight times as much protein per kilogram as milk (which is mostly water) and roughly four times as much protein as beans.[6] The

[1] There are many other greenhouse gases, such as chlorofluorocarbons, hydrofluorocarbons, and hydrochlorofluorocarbons, but carbon dioxide, methane, and nitrous oxides account for almost all agricultural greenhouse gas emissions.

[2] IPCC (2019).

[3] Allen et al. (2018), Cain et al. (2019), Lynch et al. (2020, b), Dan Vimont, personal communication (2021).

[4] Liu et al. (2021).

[5] Drewnowski et al. (2015).

[6] Johns Hopkins Medicine (2019).

bioavailability of animal proteins is also greater than that of vegetable proteins.[7] Some studies measure greenhouse gas emissions per unit of land or per animal, but those measures do not directly relate to our ultimate purpose when raising livestock, i.e. milk or meat production.

Second, what do we mean by modern intensive livestock production, and what system or systems are we comparing it to? Table 4.1 shows some of the major differences between modern intensive livestock production and traditional extensive livestock systems.

In Table 4.1 the distinctions between intensive and extensive livestock production appear to be clearly delineated, but in the real world, many farms have some characteristics of each. This is especially true for beef production in North America

Table 4.1 Key differences between intensive and extensive livestock production

Practice	Intensive livestock	Extensive livestock
Breeding	Artificial insemination often used in dairy cattle. May use other technologies, including embryo transfer.	Relatively little use of off-farm breeding stock or artificial insemination
Housing	Animals housed or kept in specially designed outdoor holding areas, environment (ventilation, temperature, lighting) may be controlled.	Animals kept outdoors, sometimes with access to basic shelter but no environmental control.
Feed	All feed brought to animals, diets formulated to meet nutrient requirements for growth or milk production. Diets often contain grains, legumes or by-product feeds.	Animals get most or all of their feed by grazing and / or foraging. Animals may migrate or be moved to different pastures.
Medication	Animals are vaccinated and treated to prevent parasites. In the event of a disease outbreak, animals are treated with appropriate medication.	Greater reliance on vaccination to control disease as animals are not necessarily routinely handled.
Supplements and growth promoters	Animals are provided mineral supplements and may receive hormones, subtherapeutic antibiotics and other growth promoters, depending on the production system.	Mineral supplements often used, but few growth-promoting technologies are employed.
Manure	Manure is removed from housing and stored, often in anaerobic lagoons. Manure may or may not be treated before it is applied to land.	Manure is deposited directly onto pasture as animals graze.
Size	Varies from small to very large, (thousands of cattle or pigs, or tens of thousands of chickens) managed at one facility.	Often small, with fewer than 10 cattle, sheep, or pigs, or fewer than 100 chickens; but may be large (hundreds of cattle, thousands of chickens) depending on region and systems.

[7] McAuliffe et al. (2018).

and parts of Europe, where the cows and young calves are kept on pasture but are vaccinated and dewormed and receive mineral supplements. Then, after the calves are weaned at around 6 months of age, they may go directly to an intensive feedlot, or they may continue to graze for a few months (often with hormone implants used to improve their growth rate) and only spend the last 4–6 months of their life in an intensive feedlot. Similarly, in the U.S. many dairy cows kept on pasture are artificially inseminated; while in intensive housed dairy farms, some cows are served by the farm bull if artificial insemination does not succeed on the first or second try. In the beef industry, only a small minority of beef cows are artificially inseminated, even if their offspring will be finished in a feedlot. Because this mix of extensive and intensive production practices is common for ruminants in livestock production systems that are relatively intensive compared to the global average, the Food and Agriculture Organization (FAO) of the United Nations uses the terms "grassland" and "mixed" systems to differentiate between the fully grass-based production practices common in Sub-Saharan Africa and parts of Asia and South America, and the more intensive but still partially grass-based livestock production systems characteristic of ruminant livestock in Western Europe and North America.[8]

In contrast, most pigs and poultry in intensive livestock systems in Western Europe and North America fully conform to the intensive livestock paradigm throughout their lives. Nevertheless, in extensive or pasture-based systems, pigs and poultry still receive significant amounts of added grain and food by-products, as they are not ruminants and cannot survive on a purely forage diet. Thus, for pork and poultry production the FAO distinguishes between "industrial" and "backyard" (often seen in low to middle-income countries) systems, and also recognizes an intermediate category for pig production.[9]

This chapter will focus primarily on U.S. livestock production. In the U.S., even pasture-fed beef cattle that are not fed any grain or by-product feeds often receive mineral supplements, are vaccinated against disease, and may have access to housing during extreme weather. The critical difference between intensive and grass-based beef production systems in the U.S. is that beef animals in intensive systems spend the last 4–8 months of their lives in feedlots, where they are fed a primarily grain-based diet and may receive growth-promoting technologies including steroid hormones, subtherapeutic antibiotics, and beta-agonists (feed additives that promote muscle growth and reduce fat accumulation).[10] In contrast, grass-finished beef cattle are fed forage-based diets without supplementary feed until they are ready to be slaughtered, and growth promoting technologies are not usually used.[11]

In intensive dairy systems in the U.S., most lactating cows are housed in barns and are fed a diet that typically includes a mix of fresh or dried alfalfa and grass;

[8] Opio et al. (2013).

[9] MacLeod et al. (2013).

[10] Capper (2012), Felix (2017).

[11] There are several hormone implants that work well in grass-fed cattle; however many grass-finished cattle in the U.S. and western Europe are marketed under labels that prohibit their use.

maize, alfalfa or grass silage; grains; legumes; and by-products from human food and fiber production that cannot or will not be eaten by humans (e.g. dried distiller grains, citrus pulp, or bakery waste).[12] Dairy cattle are not given technologies to improve their body growth, but, if permitted by the market, they may be given a growth hormone, recombinant Bovine Somatotropin (rBST), which stimulates greater milk production.[13] By contrast, dairy cows in grazing systems have access to pasture whenever conditions are suitable for grazing (note that in most regions this does not mean all year round), and usually don't receive rBST. For the purposes of this chapter, we will compare grass-based dairy cows that do not consume any grain supplements with cows in intensive systems. However, in practice the delineation between "intensive" and "extensive" is not clearly marked, and many pastured dairy cows receive grain or other feed supplements.[14]

This chapter will focus primarily on the value of modern intensive production practices and technologies for reducing greenhouse gas emissions associated with U.S. beef and dairy production. We focus on ruminants because the differences in greenhouse gas emissions intensity between intensive and grass-finished beef and dairy production systems are far greater than the emissions intensity differences between industrial and backyard systems for pigs and chickens. In addition, the greenhouse gas emissions of ruminant livestock have received far more media, consumer and policy-maker attention than those of monogastric livestock, primarily because rumination inevitably leads to considerable CH_4 emissions. It is therefore not surprising that there is little research on the greenhouse gas emissions of non-intensive pork and poultry in the U.S.[15] However, many of the same breeding, feeding, and housing practices that have reduced the greenhouse gas emissions intensity of beef and dairy have resulted in similar increases in efficiency for pork and chicken, and we will touch on those briefly.

---

[12] Food wastes and by-products fed to dairy cows range from standard grains such as oats and wheat that do not meet standards for human use to fruit pulp and other food by-products (Walker, 2000).

[13] Dairy cows may receive therapeutic antibiotics to treat disease, but their milk cannot be used for human consumption while the animal is being treated and for a period after treatment ends to ensure that the medicines are not present in milk consumed by people. The growth hormone rBST is used in the U.S. (marketed under the brand name Posilac™), but is not approved for use in many other countries, including Canada, the European Union, and New Zealand.

[14] There is a lot of variation among dairy farms that practice grazing. The authors know of a farm that uses rBST and also puts the cows on pasture when the weather permits, but that is an unusual case. Many grazing dairies sell their milk into the organic market, which requires that the cows get at least 30% of their nutrition from pasture when the weather permits and prohibits use of rBST. Some organic farms just meet the minimum grazing requirements; others may sell their milk or cheese under a grass-fed (no grain) label for an additional premium; and many fall somewhere between those extremes.

[15] The differences in emissions intensity for poultry and pig production are much greater between regions within the industrial or backyard systems than between the two systems MacLeod et al. (2013).

4.2 Regional Differences

As discussed in Chap. 2, on average, a kilogram of milk or beef produced in Europe or North America generates far lower greenhouse gas emissions than the same amount of milk or beef produced in South America, South Asia, or Africa (Figs. 4.1 and 4.2).[16] In Europe and North America, modern intensive production systems are commonly found within the dairy and beef sectors, while extensive grass-based production is exhibited in the majority of cattle systems in South America, South Asia, and Sub-Saharan Africa.

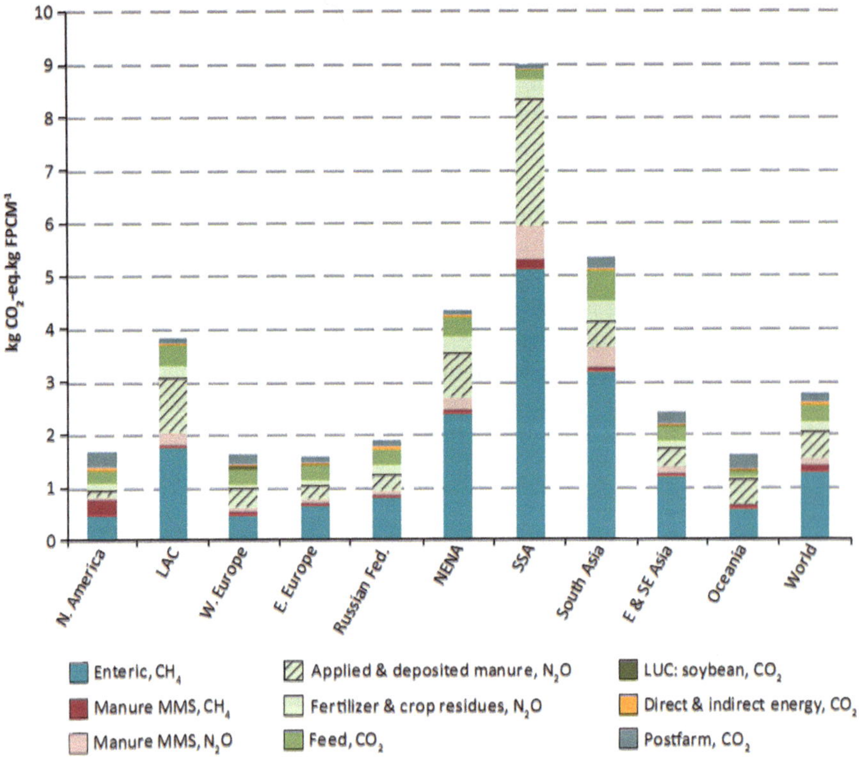

Fig. 4.1 Regional variation in emission intensities for cow milk as measured by the FAO's GLEAM model (Opio et al., 2013). FPCM refers to milk with a standardized amount of butterfat and protein (fat and protein corrected milk). LAC refers to Latin America and the Caribbean; NENA to Near East & North Africa; SSA to Sub-Saharan Africa; and Oceania to New Zealand and Australia. MMS stands for manure management system, and LUC stands for land use change (e.g., conversion of forest to cropland or pasture)

[16] Opio et al. (2013).

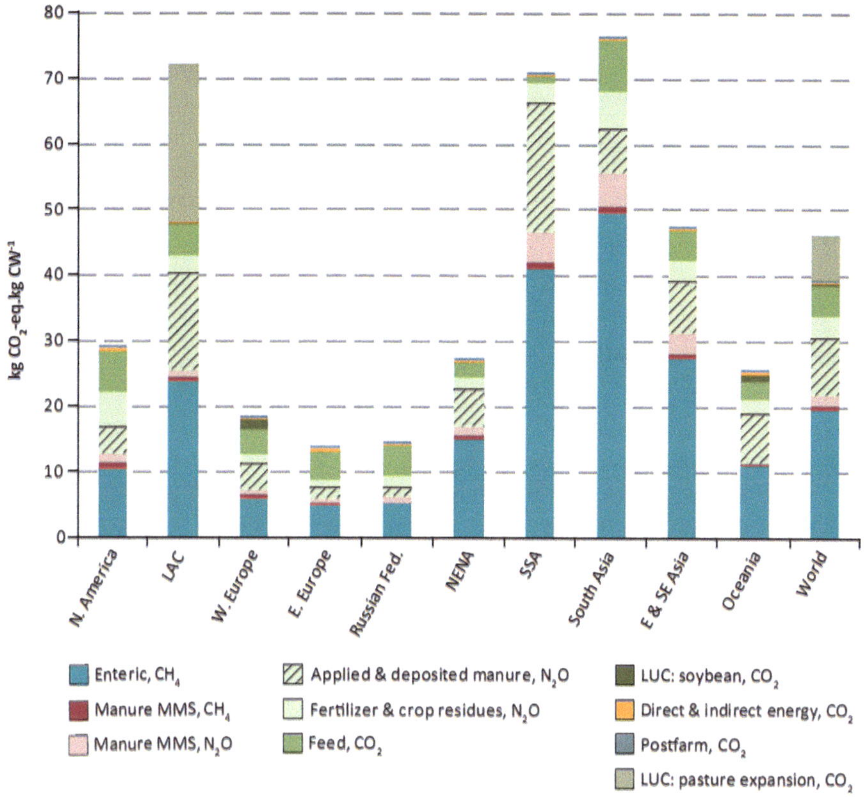

Fig. 4.2 Regional variation in emission intensities for beef as measured by the FAO's GLEAM model (Opio et al., 2013). CW refers to carcass weight, MMS to manure management systems, and LUC to land use change (e.g., conversion of forest to cropland or pasture). LAC refers to Latin America and the Caribbean; NENA to Near East & North Africa; SSA to Sub-Saharan Africa; and Oceania to New Zealand and Australia

4.3 Temporal Differences – The Dairy Case

The regional variations in greenhouse gas emissions intensity are stark, but are they necessarily due to clear differences in production system, or are they more nuanced? One way to examine the impact of changes in production system, while controlling the effects of geographic influence (and to some degree, markets and culture) is to examine changes in greenhouse gas emissions over time in the same region. In the 1940s dairy herds in the U.S. contained only a few cows and relied primarily on pasture and hay for feed. Inorganic fertilizers and other cropping inputs were not available, and neither antibiotics nor artificial insemination had yet been commercialized. To some degree, these systems faced many of the same challenges as contemporary extensive, grass-based dairy systems in sub-Saharan Africa or South Asia. The average U.S. cow in 1944 produced 2074 kg of milk per year, and the estimated greenhouse gas emissions intensity for that milk was 3.66 kg CO_2eq per kilogram of milk.[17]

In 2007, the average U.S. dairy cow produced 9193 kg of milk per year, with an emissions intensity of 1.35 kg CO_2eq per kilogram of milk; and in 2017 her descendent produced 10,406 kg of milk per year, with a greenhouse gas emissions intensity almost 20% lower than that for milk in 2007.[18] As a result of these efficiency gains, the dairy sector in 2017 produced almost 25% more milk than 10 years earlier, but only increased greenhouse gas emissions by 1%.

4.4 Efficiency Gains

So how did modern intensive dairy production achieve these efficiency gains? Let's look at four factors: genetics, nutrition, technology use, and housing.

Genetics play an important role in how much milk a cow can produce and how efficient she is at converting feed into milk. Farmers have been breeding cows for greater milk production for hundreds of years but relying on natural service means that the pace of genetic improvement is relatively slow, with a gene pool restricted to a small number of farm bulls. By contrast, artificial insemination allows a farm to bring in top genetics from anywhere in the country – or even from abroad. Since the introduction and widespread use of artificial insemination significant gains have been made in milk production efficiency, yet as wide variation in milk yield still exists, there is considerable potential for further genetic improvement.[19]

Milk yield directly affects the quantity of greenhouse gas emissions produced per unit of milk. A proportion of the nutrients within dairy cow feed are used to maintain bodily functions and health (maintenance nutrients), a further proportion is used for production – either growth, pregnancy or lactation, and some are excreted as waste in the form of feces, urine, and methane. The quantity of nutrients (and therefore feed) required for maintenance varies according to bodyweight and activity level. The average dairy cow in 2017 weighed 748 kilograms, while the average dairy cow in 1944 weighed only about 518 kilograms, therefore the modern dairy cow needs almost one and a half times as many nutrients just to keep her alive and functioning. On the other hand, a dairy cow in 2017 produced five times as much milk as the average cow in 1944, so a much larger percentage of the modern cow's feed consumption goes toward making milk, and a smaller percentage goes toward keeping the cow alive and making methane and manure. This is called the "dilution of maintenance" effect – simply put, as milk yield increases, the nutrients required

[17] Capper et al. (2009).

[18] Capper et al. (2009), Capper and Cady (2020). Because of a combination of the estimate of the global warming potential of methane changing between 2009 and 2020, and other adjustments to the model calculations, there is not a direct comparison between the greenhouse gas emissions intensities of 1944 and 2017.

[19] Capper and Cady (2020).

for maintenance (the fixed costs of dairy production) are diluted over more units of production. Proportionally, a cow with a higher milk yield requires fewer nutrients, and less feed, water and land per kg of milk produced, and has lower greenhouse gas emissions per unit of milk. The dilution of maintenance effect also means that we need fewer modern high-yielding cows to produce a set quantity of milk, therefore the entire herd requires fewer resources and has lower greenhouse gas emissions than lower-producing herds.[20]

Genetics alone do not assure milk production however – dairy cows still need well-balanced diets of high-quality feed to produce a lot of milk. In traditional extensive pasture systems, the quality of the feed is often highly variable and difficult to manage. Some plants are easily digestible and high in carbohydrates and proteins, while others are high in lignin or other fibers that are less digestible. The age of the plant also plays a role. Grasses and legumes tend to have highest nutritional quality when they are young, and most quickly lose quality (simple carbohydrate and protein content) and become more woody when they start to flower. Once the plants become woody, it is more difficult for the microorganisms in the rumen to break them down, so cows eating fibrous, low-quality forage produce less milk and also tend to emit more methane per unit of meat or milk produced than cows eating a high-quality diet.[21] In intensive dairy systems forages are harvested when their quality is high, and are then fermented (silage) or dried (hay) to preserve nutritional quality. In addition, forages are mixed with grains and concentrates that are high in energy and protein to provide cows the optimum balanced diet for milk production. In traditional pasture-based systems forage quality may be high only for a short time during the year, and cattle have a greater maintenance nutrient requirement because of their activity while grazing.

Given the improvements in genetics and diet, cows in modern intensive U.S. dairy systems can produce large quantities of milk with far lower greenhouse gas emission intensities than cows in extensive grass-based systems. These efficiency gains can be boosted even further with use of performance-enhancing technologies. In the mid-1990s Monsanto began selling recombinant bovine somatotropin (rBST) to dairy farmers in the United States under the brand name Posilac™. This product is an artificial hormone that mimics a growth hormone naturally produced in the pituitary gland of ruminants and many other animals, including humans. Research dating back to early in the twentieth century showed that injecting cows with somatotropin increased their milk production.[22] However, the recombinant technology that made rBST production commercially feasible starting in the 1990s was controversial, and many consumers and some scientists had concerns about its potential impacts on animal welfare and human health.[23] The Food and Drug Administration in the U.S. concluded that rBST did not pose a danger to human

[20] Capper and Bauman (2013).

[21] Beauchemin et al. (2020), Dong et al. (2015).

[22] Lamas et al. (2019).

[23] U.S. Food and Drug Administration (2021).

health or the environment. At present, rbST has not been registered for use in the European Union and Canada, but both regions allow imports of dairy products from cows supplemented with rBST.[24]

A dairy cow supplemented with rBST produces, on average, an extra 4.54 kg (10 lb) of milk per day of lactation. A research study modeling the effect of supplementing dairy cows with rBST reported that the greenhouse gas emissions intensity of milk could be reduced by an additional 8% over and above the already low emissions intensity of milk from intensive U.S. dairy farms.[25] This reduction in environmental impact is impressive, but would need to be balanced against the aforementioned consumer concerns over hormone use in dairy production, even if these concerns do not have a scientific evidence basis.

Dairy cow health and welfare have obvious ethical implications in that there is a moral imperative to make sure that cattle are well cared for, but also has a significant impact on their climate footprint. The maintenance nutrient requirement is affected by environmental conditions as well as by animal size and activity. When temperatures are below 4 °C or above 24 °C cattle have to expend additional energy to maintain a healthy body temperature. When the weather is hot, cows tend to eat less because their digestive process generates a lot of body heat, so reducing feed intake helps keep them from overheating. As a result, under hot humid conditions milk production decreases, while under cold conditions cows need to eat more feed just to maintain their body temperature and cannot partition so many nutrients into milk production. Cows in intensive modern production systems are typically protected from temperature extremes – barns provide wind shelter in winter and often have fans and sprinkler systems to cool cows in hot weather.[26] By contrast, cows in extensive outdoor systems have little or no protection from weather extremes, and particularly in hot climates this lack of protection contributes to lower milk production and higher greenhouse gas emission intensities. There are few papers in the literature that have directly quantified the impacts of dairy cow health and welfare on greenhouse gas emissions, although some studies exist on individual diseases, e.g. mastitis.[27] Nevertheless, the improvements in veterinary treatments, technologies, and surveillance over past decades means that having tools to quickly and effectively treat a range of health conditions for dairy cows undoubtedly contributes to the efficiency, productivity, and low greenhouse gas emissions of modern dairy farms.

[24] Lamas et al. (2019), Sechen (2013), U.S. Food and Drug Administration (2021).

[25] Capper et al. (2008).

[26] Stull et al. (2008), Van Os (2019). If humidity is high, dairy cows can experience heat stress at temperatures above 21 °C.

[27] Mostert et al. (2019), Özkan Gülzari et al. (2018).

4.5 Beef

Genetic improvements, high-quality feed, and performance-enhancing technologies have also greatly increased the efficiency of modern intensive beef production and so reduced its emissions intensity. Capper (2012) modeled the effects of beef cattle average daily gain increasing from 0.68 kg/d in 1977 to 1.18 kg/d in 2007, which cut the number of days required to raise an animal from birth to slaughter from 609 to 485. Together with an increase in average slaughter weights (468 kg in 1977 vs. 607 kg in 2007) and improved crop yields, these productivity gains resulted in 19% less feed, 33% less land and 12% less water being required to produce one kg of beef in 2007, and a 16% reduction in greenhouse gas emissions per kg of beef.[28] Similar results were reported for the Canadian beef industry in 1981 compared to 2011, with one kg of beef produced in 2011 requiring 29% fewer cattle, 24% less land, 14% lower methane emissions, 15% lower nitrous oxide emissions and a 12% decrease in carbon dioxide emissions from fossil fuel use. In aggregate, this led to a 14% reduction in total greenhouse gas emissions per kg of beef.[29]

The most commonly used performance-enhancing technologies in U.S. beef production include ionophores, in-feed hormones, beta-adrenergic agonists and hormone implants.[30] Each of these technologies has a different mode of action, yet all work to improve efficiency and therefore reduce greenhouse gas emissions intensity by improving average daily gain and/or meat yield.[31] A model comparing conventional (feedlot finishing with performance-enhancing technologies), natural (feedlot finishing without performance-enhancing technologies), and grass-finished beef production (without performance-enhancing technologies) in the U.S. found that the conventional approach reduced greenhouse gas emissions per kilogram of beef by 40% compared to the grass-fed system, and by 15% compared to the natural system.[32] In this comparison the major difference between conventional and natural beef production was that the intensive cattle received growth-promoting supplements (where appropriate to the cattle and approved by the FDA), including hormone implants, in-feed hormones and beta agonists, whereas the natural beef cattle did not receive any performance-enhancing supplements.[33] In both systems the animals spent the final 4–8 months of their lives in a feedlot, where they were fed maize, soybean meal, and alfalfa hay. Grass-finished cattle did not receive performance-enhancing technologies or any supplemented feed.

---

[28] Capper (2012).

[29] Legesse et al. (2016).

[30] Strydom (2016).

[31] Johnson et al. (2013).

[32] Capper (2012)

[33] The antibiotics fed to stimulate growth are ionophores, a class of antibiotics not used for humans. Thus, their use for beef production is not expected to result in antibiotic resistance affecting humans.

As Fig. 4.3 shows, in the conventional system where the cattle were finished on grain and supplemented with performance-enhancing technologies they grew both faster and larger than the cattle in the natural system. The average carcass weight of grass-fed beef cattle was lower than either conventional or natural cattle, even though they were roughly 50% older than the conventionally raised cattle. Although the smaller grass-fed cattle ate less feed and produced less manure per day than the larger conventional cattle, because they lived longer they ate more feed and produced more manure over their lifetime than their conventional counterparts. They also emitted more enteric methane over their lifetime, both as a function of their older age at slaughter and because cattle on a pure grass diet typically produce more enteric methane than cattle supplemented with concentrates or grains.[34] Similar effects of finishing system and technology use were reported in live (not modeled) cattle experiments.[35]

Removing performance-enhancing technologies (PET) from U.S. cattle systems in a modeling simulation was shown to reduce slaughter weight by 53 kg and increase feed, land and water use by 11%, 10% and 4% per kg of beef respectively.[36] Total greenhouse gas emissions per kg of beef produced without PET were increased by 10%, and economic costs of production were increased by 8%. Consequently, the authors predicted that if PET were withdrawn completely, U.S. beef production would decrease by 17.1% by 2023, with the deficit compensated for by increases in beef production in Canada, Brazil, Argentina and Australia. Although removing performance-enhancing technologies from U.S. beef production has been a topic of

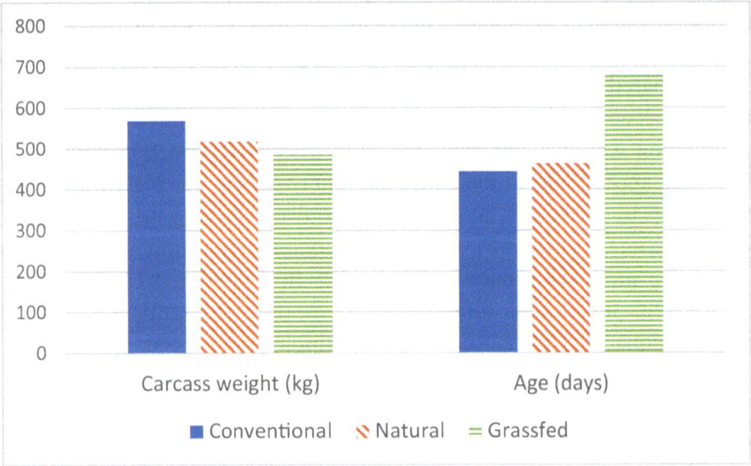

Fig. 4.3 Average carcass weight in kilograms and age at slaughter of beef cattle raised in conventional, natural, and grass-based systems. (Data from Capper, 2012)

[34] Beauchemin et al. (2020).

[35] Basarab et al. (2012), Stackhouse et al. (2012), Webb et al. (2017).

[36] Capper and Hayes (2012).

significant discussion over the past decade, the environmental and economic benefits of performance-enhancing technologies use would seem to render it an unwise move, yet these products must be made more acceptable to consumers, now and in future.

4.6 Poultry and Pork

Both the popular press and the scientific literature have focused more on greenhouse gas emissions from cattle than on emissions from pork and poultry production, primarily because the emissions intensity of pork and chicken is considerably lower than that of beef. As pigs and chickens are not ruminants, their primary greenhouse gas emissions come from feed production and manure management; therefore the differences between intensive and backyard systems are smaller than those of dairy or beef cattle systems.

As a result of improvements in genetics, feed quality, housing and health, pigs in intensive production systems are roughly 1.8 times more efficient at converting feed to meat than pigs in backyard systems, and about twice as efficient as in the 1960s.[37] Within U.S. swine production, the number of pigs marketed increased from 87.6 million in 1959 to 112.6 million in 2009 (a 29% increase), from a breeding herd that decreased in size by 39% over the same time period, conferring reductions in feed, land, water and greenhouse gas emissions per kg of pork of 67%, 41%, 22% and 35%, respectively.[38] However, differences between systems in feed production result in similar emission intensities for backyard and industrial pork production systems on a global basis (Fig. 4.4). In particular, intensive pork production in Western Europe and Latin America relies on soybeans produced in Brazil, which are associated with high carbon dioxide emissions from Amazonian deforestation, while backyard production relies on local grains and food waste, which have relatively low emission factors.[39]

Likewise, although there are regional differences in pig production's greenhouse gas emissions intensity (Fig. 4.5), these differences are smaller than the regional differences for beef and dairy, and they do not correlate closely with whether intensive or backyard production predominates in the region. Western Europe and North America have the world's most intensive pork systems, with relatively little non-intensive production compared to low- and middle-income countries. Western Europe's emission intensity is higher than that of several regions dominated by backyard production, while North America has the lowest emission intensity of all the regions evaluated. A closer look reveals that reliance on soybeans imported from

[37] The feed conversion ratio for industrial pork is roughly 2.75, with little regional variation, while for backyard pork production it ranges from just over 4 to 6, with a world average of 5 kg of feed dry matter needed to produce 1 kg live weight (Hume et al., 2011; MacLeod et al., 2013).

[38] Cady et al. (2013).

[39] MacLeod et al. (2013), Wiedemann et al. (2016).

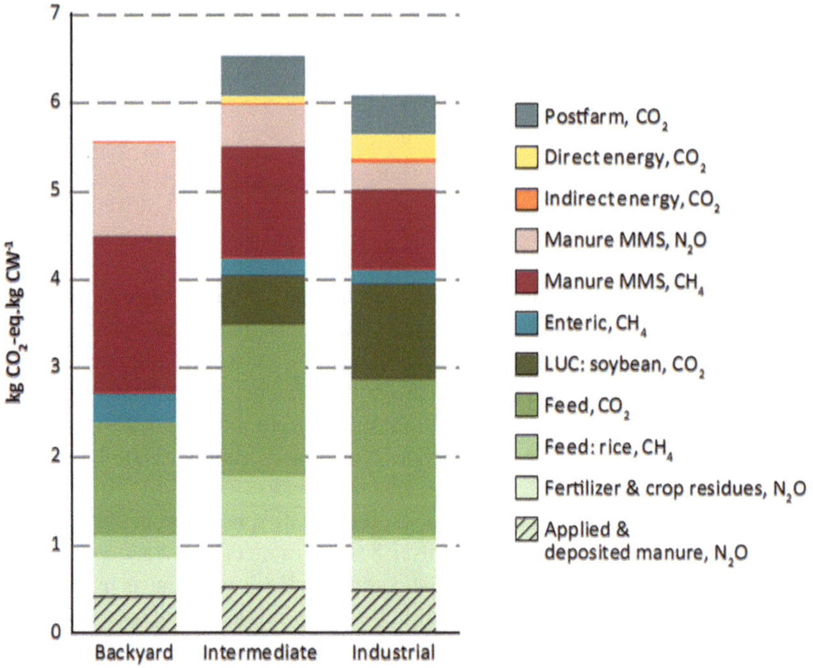

Fig. 4.4 Average global emission intensities of three different pork production systems as measured by the FAO's GLEAM model (MacLeod et al., 2013). CW stands for carcass weight, MMS stands for manure management system, and LUC stands for land use change from forest to cropland

Latin America accounts for the relatively high emission intensity of intensive pork from Western Europe. Although North America does not rely on soybean imports, its manure methane emissions are relatively higher than those in Western Europe, indicating potential for further emissions reductions through improved manure management practices such as biodigestion or methane capture and flaring.[40]

Like beef and pork, intensive poultry production has made huge strides in increasing feed efficiency. For example, from 1985 to 2010 the feed conversion ratio of meat chickens decreased from 2.3 to 1.5. In other words, in 1985 it took 5.6 kg of feed to raise a chicken weighing 2.44 kg, but in 2010 it only took 3.7 kg of feed to raise the same size bird. As previously discussed with reference to other livestock, these efficiency gains are due to advances in genetics, nutrition, housing, and disease prevention and treatment.[41] Productivity gains made by the U.S. egg industry have followed the same trends over the five decades from 1960 to 2010, with lesser resource use per ton of eggs (reductions in pullet numbers of 22%, feed

[40] MacLeod et al. (2013).

[41] Hume et al. (2011), Sell-Kubiak et al. (2017), Siegel (2014).

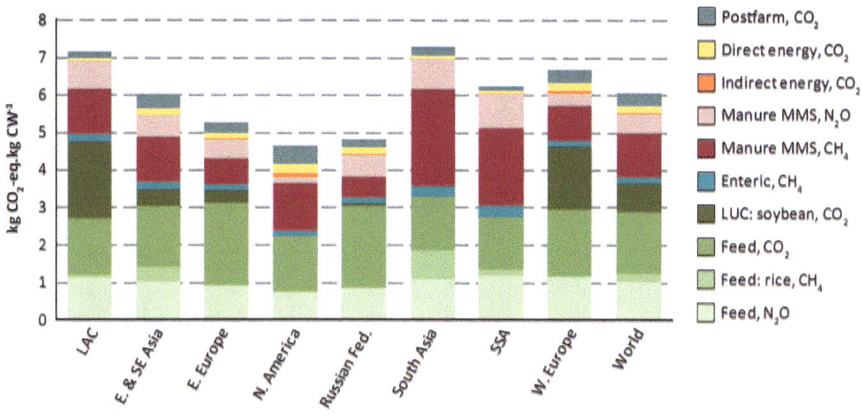

Fig. 4.5 Pork production emission intensities by region as measured by the FAO's GLEAM model (MacLeod et al., 2013). CW stands for carcass weight; MMS stands for manure management system; LUC stands for land use change from forest to cropland; LAC refers to Latin America and the Caribbean; and SSA refers to Sub-Saharan Africa

use of 42% and water use of 32%) and a 71% reduction in greenhouse gas emissions.[42]

Feed efficiency in traditional poultry systems remains low compared to that in intensive production: on average, backyard poultry need about three times more feed than intensive chickens to produce the same amount of meat or eggs.[43] Nevertheless, as shown in Fig. 4.6, the overall greenhouse gas emission intensity is only slightly lower for industrial meat and egg production than for extensive systems, largely because intensive poultry production everywhere except North America often relies heavily on soybean imports from Brazil and Argentina. Although the Food and Agriculture Organization found relatively small differences in emission intensity between modern intensive and traditional backyard pork and poultry systems, those findings were based on years when soybean imports from Brazil and Argentina were high.[44] During those same years considerable amounts of deforestation was occurring in both countries, along with a significant expansion of the amount of land used for soybean production. When feed for modern intensive pork and poultry production is produced without significant land use change, the emissions intensity is lower than for traditional production systems.

---

[42] Pelletier et al. (2014).

[43] MacLeod et al. (2013).

[44] See Appendix C (MacLeod et al., 2013)

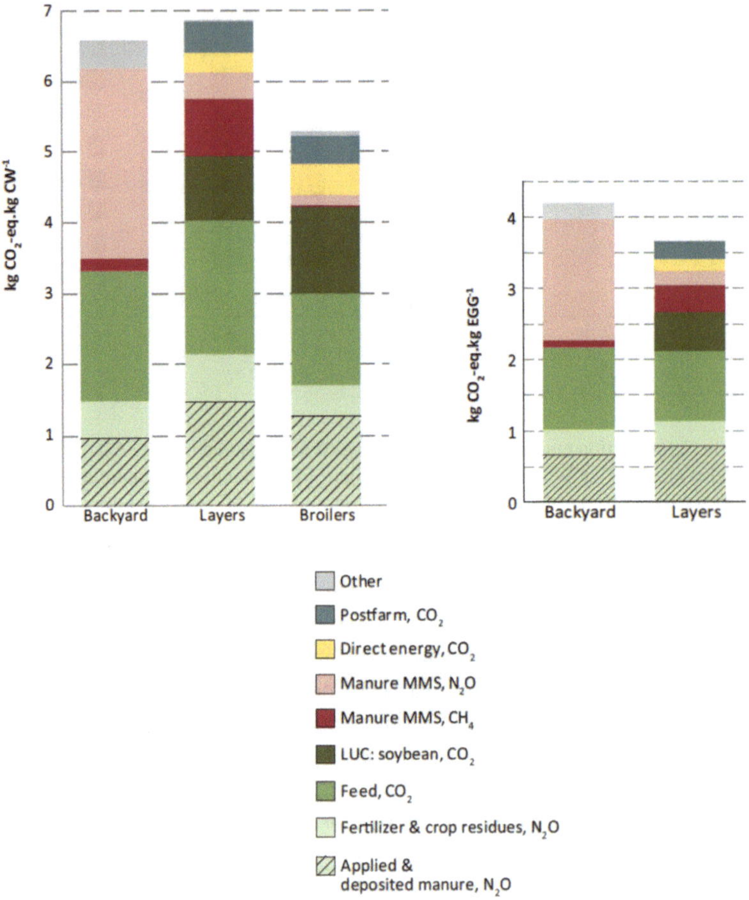

Fig. 4.6 Chicken meat emission intensity (left) and egg emission intensity (right) as measured by the FAO's GLEAM model. "Other" emissions include methane from rice production in backyard systems and indirect CO_2 emissions from feed inputs in industrial production. LUC stands for land use change, MMS stands for manure management system, and CW stands for carcass weight. (MacLeod et al., 2013)

4.7 Conclusion

Modern intensive production methods have achieved remarkable gains in efficiency through improved nutrition, genetics, management, health and technology use; and these gains have translated into reduced resource use and greenhouse gas emissions per unit of milk, meat or eggs produced, regardless of global region. Regions of the world that use intensive production techniques have much lower emission intensities from beef and dairy than regions that rely on traditional extensive grass-based production. Intensive ruminant production systems also need less land to produce the same amount of milk and meat as traditional grass-based systems, creating the potential for carbon sequestration through reforestation.[45] As the demand for nutrient-rich dairy and meat products grows with a rising world population, intensive livestock production techniques will play a critical role in curbing the greenhouse gas emissions of the livestock sector.

This does not mean that extensive production systems should be abandoned – on the contrary, considerable areas of land exist where extensive production is ideally suited to the resources, climate and livestock; and many smallholder systems in low- or middle-income countries rely on extensive livestock systems to provide a livelihood. Nevertheless, the opportunity to improve productivity exists in every livestock system worldwide. The challenge is to communicate the management practices, tools and technologies that will make this a reality, while taking note of resource, climate and cultural restrictions.

Bibliography

Allen, M. R., Shine, K. P., Fuglestvedt, J. S., Millar, R. J., Cain, M., Frame, D. J., & Macey, A. H. (2018). A solution to the misrepresentations of CO2-equivalent emissions of short-lived climate pollutants under ambitious mitigation. *Npj Climate and Atmospheric Science, 1*(1), 16. https://doi.org/10.1038/s41612-018-0026-8

Basarab, J., Baron, V., López-Campos, Ó., Aalhus, J., Haugen-Kozyra, K., & Okine, E. (2012). Greenhouse gas emissions from calf- and yearling-fed beef production systems, with and without the use of growth promotants. *Animals, 2*(2), 195–220. https://doi.org/10.3390/ani2020195

Beauchemin, K. A., Ungerfeld, E. M., Eckard, R. J., & Wang, M. (2020). Review: Fifty years of research on rumen methanogenesis: Lessons learned and future challenges for mitigation. *Animal, 14*, s2–s16. https://doi.org/10.1017/S1751731119003100

Cady, R. A., Boyd, G., Wittig, L., Bryan, G., Holden, P. J., Sutton, A. L., & Anderson, D. (2013). A 50-year comparison of the environmental impact and resource use of the U.S. swine herd: 1959 vs. 2009. *Proceedings of the ADSA-ASAS Joint Annual Meeting*.

Cain, M., Lynch, J., Allen, M. R., Fuglestvedt, J. S., Frame, D. J., & Macey, A. H. (2019). Improved calculation of warming-equivalent emissions for short-lived climate pollutants. *Npj Climate and Atmospheric Science, 2*(1), 29. https://doi.org/10.1038/s41612-019-0086-4

Capper, J. L. (2012). Is the grass always greener? Comparing the environmental impact of conventional, natural and grass-fed beef production systems. *Animals, 2*(4), 127–143. https://doi.org/10.3390/ani2020127

Capper, J. L., & Bauman, D. E. (2013). The role of productivity in improving the environmental sustainability of ruminant production systems. *Annual Review of Animal Biosciences, 1*(1), 469–489. https://doi.org/10.1146/annurev-animal-031412-103727

Capper, J. L., & Cady, R. A. (2020). The effects of improved performance in the U.S. dairy cattle industry on environmental impacts. between 2007 and 2017. *Journal of Animal Science, 98*(1), skz291. https://doi.org/10.1093/jas/skz291

Capper, J. L., & Hayes, D. J. (2012). The environmental and economic impact of removing growth-enhancing technologies from U.S. beef production 1. *Journal of Animal Science, 90*(10), 3527–3537. https://doi.org/10.2527/jas.2011-4870

[45] Garnett et al. (2017).

Capper, J. L., Castaneda-Gutierrez, E., Cady, R. A., & Bauman, D. E. (2008). The environmental impact of recombinant bovine somatotropin (rbST) use in dairy production. *Proceedings of the National Academy of Sciences, 105*(28), 9668–9673. https://doi.org/10.1073/pnas.0802446105

Capper, J. L., Cady, R. A., & Bauman, D. E. (2009). The environmental impact of dairy production: 1944 compared with 2007. *Journal of Animal Science, 87*(6), 2160–2167. https://doi.org/10.2527/jas.2009-1781

Dong, L. F., Ferris, C. P., McDowell, D. A., & Yan, T. (2015). Effects of diet forage proportion on maintenance energy requirement and the efficiency of metabolizable energy use for lactation by lactating dairy cows. *Journal of Dairy Science, 98*(12), 8846–8855. https://doi.org/10.3168/jds.2015-9465

Drewnowski, A., Rehm, C. D., Martin, A., Verger, E. O., Voinnesson, M., & Imbert, P. (2015). Energy and nutrient density of foods in relation to their carbon footprint. *The American Journal of Clinical Nutrition, 101*(1), 184–191. https://doi.org/10.3945/ajcn.114.092486

Felix, T. (2017). *Ionophores: A technology to improve cattle efficiency*. Penn State Extension. https://extension.psu.edu/ionophores-a-technology-to-improve-cattle-efficiency

Garnett, T., Godde, C., Muller, A., Röös, E., Smith, P., & de Boer, I. (2017). *Ruminating on cattle, grazing systems, methane, nitrous oxide, the soil carbon sequestration question – and what it all means for greenhouse gas emissions*, p. 127.

Hume, D. A., Whitelaw, C. B. A., & Archibald, A. L. (2011). The future of animal production: Improving productivity and sustainability. *The Journal of Agricultural Science, 149*(S1), 9–16. https://doi.org/10.1017/S0021859610001188

IPCC. (2019). *2019 refinement to the 2006 IPCC guidelines for National Greenhouse gas inventories. Volume 4.* Agriculture, Forestry and Other Land Use. Intergovernmental Panel on Climate Change. https://www.ipcc-nggip.iges.or.jp/public/2019rf/vol4.html

Johns Hopkins Medicine. (2019). *Protein content of common foods*. https://www.hopkinsmedicine.org/bariatrics/_documents/nutrition_protein_content_common_foods.pdf

Johnson, B. J., Ribeiro, F. R. B., & Beckett, J. L. (2013). Application of growth technologies in enhancing food security and sustainability. *Animal Frontiers, 3*(3), 8–13. https://doi.org/10.2527/af.2013-0018

Lamas, A., Regal, P., Vázquez, B., Miranda, J. M., Cepeda, A., & Franco, C. M. (2019). Tracing recombinant bovine somatotropin ab(use) through transcriptomics: The potential of bovine somatic cells in a multi-dose longitudinal study. *Scientific Reports, 9*(1), 4788. https://doi.org/10.1038/s41598-019-41343-6

Legesse, G., Beauchemin, K. A., Ominski, K. H., McGeough, E. J., Kroebel, R., MacDonald, D., Little, S. M., & McAllister, T. A. (2016). Greenhouse gas emissions of Canadian beef production in 1981 as compared with 2011. *Animal Production Science, 56*(3), 153. https://doi.org/10.1071/AN15386

Liu, S., Proudman, J., & Mitloehner, F. M. (2021). Rethinking methane from animal agriculture. *CABI Agriculture and Bioscience, 2*(1), 22. https://doi.org/10.1186/s43170-021-00041-y

Lynch, J., Cain, M., Pierrehumbert, R., & Allen, M. (2020). Demonstrating GWP*: A means of reporting warming-equivalent emissions that captures the contrasting impacts of short- and long-lived climate pollutants. *Environmental Research Letters, 15*(4), 044023. https://doi.org/10.1088/1748-9326/ab6d7e

Lynch, J., Garnett, T., Persson, M., Röös, E., & Reisinger, A. (2020). *Methane and the sustainability of ruminant livestock* (p. 20). Food Climate Research Network.

MacLeod, M., Gerber, P., Mottet, A., Tempio, G., Falcucci, A., Opio, C., Vellinga, T., Henderson, B., & Steinfeld, H. (2013). *Greenhouse gas emissions from pig and chicken supply chains: A global life cycle assessment*. Food and Agriculture Organization of the United Nations. http://www.fao.org/3/i3460e/i3460e00.htm

McAuliffe, G. A., Takahashi, T., & Lee, M. R. F. (2018). Framework for life cycle assessment of livestock production systems to account for the nutritional quality of final products. *Food and Energy Security, 7*(3), e00143. https://doi.org/10.1002/fes3.143

Mostert, P. F., Bokkers, E. A. M., de Boer, I. J. M., & van Middelaar, C. E. (2019). Estimating the impact of clinical mastitis in dairy cows on greenhouse gas emissions using a dynamic stochastic simulation model: A case study. *Animal, 13*(12), 2913–2921. https://doi.org/10.1017/S1751731119001393

Opio, C., Gerber, P., Mottet, A., Falcucci, A., Tempio, G., MacLeod, M., Vellinga, T., Henderson, B., & Steinfeld, H. (2013). Greenhouse gas emission from ruminant supply chains: A global life cycle assessment. *Food and Agriculture Organization of the United Nations*. http://www.fao.org/3/i3461e/i3461e00.htm

Özkan Gülzari, Ş., Vosough Ahmadi, B., & Stott, A. W. (2018). Impact of subclinical mastitis on greenhouse gas emissions intensity and profitability of dairy cows in Norway. *Preventive Veterinary Medicine, 150*, 19–29. https://doi.org/10.1016/j.prevetmed.2017.11.021

Pelletier, N., Ibarburu, M., & Xin, H. (2014). Comparison of the environmental footprint of the egg industry in the United States in 1960 and 2010. *Poultry Science, 93*(2), 241–255. https://doi.org/10.3382/ps.2013-03390

Sechen, S. J. (2013). *Bovine somatotropin (bST) – Possible increased use of antibiotics to treat mastitis in cows, October 30, 2013* (p. 8). U.S. Food and Drug Administration. https://www.fda.gov/media/88152/download

Sell-Kubiak, E., Wimmers, K., Reyer, H., & Szwaczkowski, T. (2017). Genetic aspects of feed efficiency and reduction of environmental footprint in broilers: A review. *Journal of Applied Genetics, 58*(4), 487–498. https://doi.org/10.1007/s13353-017-0392-7

Siegel, P. B. (2014). Evolution of the modern broiler and feed efficiency. *Annual Review of Animal Biosciences, 2*(1), 375–385. https://doi.org/10.1146/annurev-animal-022513-114132

Stackhouse, K. R., Rotz, C. A., Oltjen, J. W., & Mitloehner, F. M. (2012). Growth-promoting technologies decrease the carbon footprint, ammonia emissions, and costs of California beef production systems1. *Journal of Animal Science, 90*(12), 4656–4665. https://doi.org/10.2527/jas.2011-4654

Strydom, P. E. (2016). Performance-enhancing technologies of beef production. *Animal Frontiers, 6*(4), 22–30. https://doi.org/10.2527/af.2016-0040

Stull, C. L., Messam, L. L., Mc, V., Collar, C. A., Peterson, N. G., Castillo, A. R., Reed, B. A., Andersen, K. L., & VerBoort, W. R. (2008). Precipitation and temperature effects on mortality and lactation parameters of dairy cattle in California. *Journal of Dairy Science, 91*(12), 4579–4591. https://doi.org/10.3168/jds.2008-1215

U.S. Food and Drug Administration. (2021, April). *Bovine Somatotropin (bST)*. U.S. Food and Drug Administration. https://www.fda.gov/animal-veterinary/product-safety-information/bovine-somatotropin-bst

Van Os, J. M. C. (2019). Considerations for cooling dairy cows with water. *Veterinary Clinics of North America: Food Animal Practice, 35*(1), 157–173. https://doi.org/10.1016/j.cvfa.2018.10.009

Walker, P. (2000). The use of food waste as a feedstuff for ruminants. In *Food waste to animal feed* (pp. 185–226). Iowa State University Press. https://onlinelibrary-wiley-com.ezproxy.library.wisc.edu/doi/pdf/10.1002/9780470290217#page=184

Webb, M. J., Pendell, D. L., Harty, A. A., Salverson, R. R., Rotz, C. A., Underwood, K. R., Olson, K. C., & Blair, A. D. (2017). Influence of growth promoting technologies on animal performance, production economics, environmental impacts and carcass characteristics of beef. *Meat and Muscle Biology, 1*(3), 23–24. https://doi.org/10.22175/rmc2017.022

Wiedemann, S. G., McGahan, E. J., & Murphy, C. M. (2016). Environmental impacts and resource use from Australian pork production assessed using life-cycle assessment. 1. Greenhouse gas emissions. *Animal Production Science, 56*(9), 1418. https://doi.org/10.1071/AN15881

Chapter 5
The Limits of Efficiency

Diane Mayerfeld

Abstract Although increasing the efficiency of animal agriculture can help reduce livestock's carbon hoofprint, the focus on efficiency ignores important issues that we cannot afford to disregard. One of these issues is the fact that the production of grain to feed livestock in industrial systems generally causes soils to lose carbon. When soil carbon fluxes are incorporated into life cycle assessments of industrial livestock, their estimated emission intensity goes up. The other major issue associated with efficient industrial livestock production is its tendency to overproduce. This overproduction results in pointless greenhouse gas emissions to produce and dispose of unwanted meat and milk, as well as unhealthy consumers and unsustainably low prices for many farmers.

Keywords Climate change · Emission intensity · Feed efficiency · Food waste · Greenhouse gas emissions · Livestock · Overproduction · Soil carbon

5.1 Introduction

Improvements in livestock genetics and in our understanding of livestock nutrition have made meat and dairy production in western Europe and North America amazingly efficient. This efficiency is one reason why agriculture is a relatively minor contributor to the overall emissions of these wealthy nations, even though our per capita consumption of livestock products is high. As a result, the UN's Food and Agriculture Organization identifies improving the efficiency of livestock production in other nations as a key strategy for reducing livestock's global greenhouse gas emissions. Even critics of animal agriculture recognize the role of efficiency in

D. Mayerfeld (✉)
Division of Extension, Agriculture Institute, University of Wisconsin, Madison, WI, USA
e-mail: dbmayerfeld@wisc.edu

D. Mayerfeld (ed.), *Our Carbon Hoofprint*, Food and Health,
https://doi.org/10.1007/978-3-031-09023-3_5

reducing the impact of livestock on climate.[1] However, there are several important issues that the focus on efficiency ignores that we cannot afford to disregard.

As we explore these blind spots it will be helpful to keep in mind that the efficiency argument has been used in response to two different challenges. One of these challenges is the concern that the climate impacts of animal agriculture are unacceptably high. The other challenge to modern, efficient agriculture is that some farmers and consumers have embraced livestock production based on managed grazing as a solution to problems associated with large-scale industrial meat and dairy production. These problems include concerns about animal welfare, bad labor conditions for farmworkers and workers at food processing facilities, and water and air pollution. Consumers seek out meat and dairy with organic, grass-fed, and humane labels that require giving livestock access to pasture when the weather allows, and consumers also buy pasture-raised meat directly from farmers without formal certifications. Although these livestock products constitute a small portion of the American and European markets for meat and dairy, the sector is growing and is seen as a threat by conventional industrial animal agriculture.[2]

So in the debate about how to address the climate impacts of agriculture livestock farmers are deeply divided about the best response, much like the version of the tragedy of the commons from the introduction where cattle, sheep, and horse farmers fought among themselves about which livestock type was most damaging to the pasture.

5.1.1 Soil Carbon

This rift between industrial and grass-based agriculture is particularly evident in how each side views the role of soil carbon in affecting climate change. Advocates for industrial livestock focus primarily on two closely related emission factors: the feed efficiency of the livestock (how much feed an animal must eat to produce a given amount of meat or milk or eggs), and how quickly the animal produces the meat. These are indeed critical factors to consider when assessing the greenhouse gas emissions of meat and dairy. But advocates of grazing insist that we also need to look at a third emissions factor: whether the soil is storing or emitting carbon.

There are good reasons to focus on feed efficiency. In the U.S. soil nitrous oxide emissions from cropland account for about 28% of overall agricultural emissions, and around 60% of that cropland is used to raise animal feed. Nitrous oxide emissions from grazed land account for approximately 17% of agricultural emissions.

[1] Steinfeld et al. 2006, pp 119–120; Garnett et al. 2017.

[2] Authors like Nicolette Hahn Niman and Michael Pollan, as well as reports on groundwater contamination in NE WI have received a lot of public attention, and conventional farmers feel demonized by these and other critiques of industrial agriculture. Dairy Business Association and Edge Dairy Farmer Cooperative 2020.

https://www.morningagclips.com/dairy-groups-push-back-on-anti-cafo-resolution/

Feed production also has indirect greenhouse gas emissions, including the fossil fuel energy used to make fertilizer and pesticides. If the emissions associated with producing livestock feed account for more than a third of agricultural greenhouse gas emissions in the U.S., then clearly increasing feed efficiency can reduce livestock's climate impact. In addition, the greater an animal's feed efficiency, presumably the less carbon is wasted as manure, which accounts for an additional 13% of agriculture's overall greenhouse gas emissions.[3]

And feed efficiency is related to time. The more efficient the animal is at converting feed to muscle, the faster it can grow and reach slaughter weight. This time dimension is particularly important for cattle and other ruminants, because the longer the animal is alive, the more time it has to emit methane as a by-product of its digestive process. In addition, animals tend to grow quickly when they are fed grain, and they also tend to produce less methane on diets that are relatively high in carbohydrates and low in fiber, such as grain. The USDA estimates that nationwide soil nitrous oxide emissions from feed production, methane from enteric emissions, and methane and nitrous oxide emissions from manure add up to about 75% of total U.S. agricultural greenhouse gas emissions, and almost all of livestock's emissions.[4] So just looking at those three greenhouse gas sources would seem to make sense.

However, this analysis ignores the amazing ability of soil to act as both a carbon sink and a carbon source. In aggregate, the U.S. Department of Agriculture (USDA) estimates that the nation's agricultural soils don't store or release much carbon, but that aggregate hides a lot of variation. Let's take a closer look at what we know about soil carbon, what we don't know, and how USDA reached its estimates.

First, soil has the potential to store a lot of carbon – globally, soils contain about 2300–2500 billion tons of carbon, compared to about 840 billion tons in the atmosphere.[5] This soil carbon pool is in constant flux with atmospheric carbon (Fig. 5.1). Plants take carbon in the form of CO_2 out of the atmosphere, convert it into carbohydrates through photosynthesis, and put it into the soil in the form of root growth, root exudates,[6] and dead plant material. Soil biology strongly influences the amount of carbon that stays in the soil. Microbes and plant matter in the soil, both living and dead, constitute an important part of soil carbon. At the same time, soil organisms also breathe, converting soil carbon back to atmospheric CO_2. Soil disturbance, like clearing land and growing crops, typically speeds up this respiration and results in loss of soil carbon. Scientists estimate that over 10,000 years agricultural activity has resulted in a loss of roughly 100 billion tons of soil carbon to the atmosphere.[7] If we can change agriculture so that it does not increase loss of soil carbon, or even

[3] U.S. Department of Agriculture (2016a, b), U.S. Environmental Protection Agency (2020).

[4] U.S. Department of Agriculture (2016a, b).

[5] Toensmeier (2016), U.S. Department of Energy Office of Science (2008). Toensmeier offers an in-depth examination of the potential for agriculture to store rather than lose soil carbon.

[6] Exudates are fluids containing carbohydrates that seep out of plant roots.

[7] Sanderman et al. (2018), Toensmeier (2016), Lal (2002).

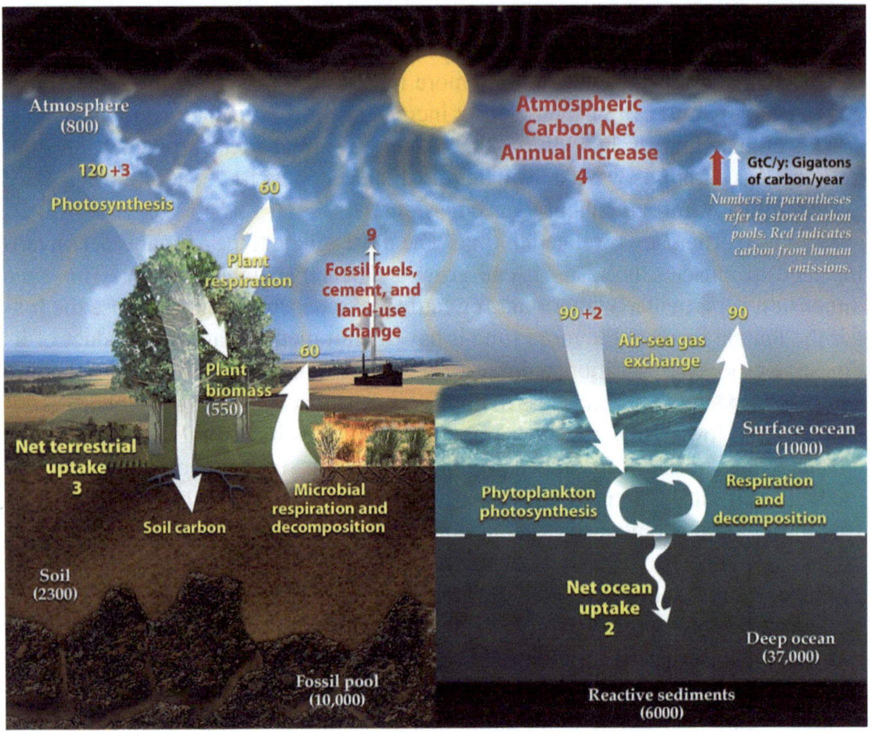

Fig. 5.1 Global carbon fluxes and stocks in 2008. Yellow numbers are natural fluxes, and red are human contributions in gigatons of carbon per year. White numbers indicate stored carbon. (U.S. Department of Energy Office of Science (2008). Diagram adapted from U.S. DOE, Biological and Environmental Research Information System. – http://earthobservatory.nasa.gov/Features/CarbonCycle/Public Domain, https://commons.wikimedia.org/w/index.php?curid=19434238)

better so it promotes soil carbon storage, then we can greatly reduce agriculture's overall climate impact.

How much carbon soil can store depends on land use and management, as well as many other factors. Fine-textured soils with a lot of clay and silt usually contain more carbon than coarse soils with a lot of sand. Cooler, drier conditions are more likely to result in carbon accumulation in the soil than warmer and wetter conditions, because soil microbes respire less when the soil is cold or dry. On the other hand, the more plant growth there is, the more carbon is available to be added to the soil, and plants tend to grow faster in warm, humid conditions.[8]

Usually the concentration of carbon in the soil is highest near the soil surface. This layer, called the A horizon, is where the greatest root mass is located, as well

[8] U.S. Department of Energy Office of Science (2008), Magdoff and van Es (2021).

as the greatest mass of soil organisms. However, carbon is also present at greater depths, though usually at lower concentrations. Because the A horizon has the highest concentration of organic matter, most researchers have focused on the top 15–30 cm of the soil when looking at the impacts of land use and management on soil carbon.[9]

This research has shown that the amount of carbon in the soil is strongly influenced by land use and by management practices. Land in perennial vegetative cover such as forest or grassland generally has higher soil carbon than land in annual crops. And within cropland, practices such as minimizing tillage and maintaining year-round ground cover result in higher amounts of soil carbon than fields that are tilled or left bare for part of the year. Researchers have also noted that the concentration of carbon in the A horizon changes relatively quickly in the first 5–20 years following a change in land use or management, but after the initial period the rate of change usually slows exponentially.[10]

The observation that the rate of change in near surface soil carbon concentrations slows dramatically over time led USDA to conclude that when looking at CO_2 flux in agricultural soils the only factor they needed to include in their model was major recent land use change. So if cropland was converted to grassland or forest, that land use change was assumed to result in CO_2 being taken out of the air and stored in the soil. Conversely, if grassland or forest was converted to cropland that change was assumed to result in CO_2 emissions. However, USDA's model assumed that after the change soil carbon under each land use would quickly reach a new equilibrium.

In turn, many researchers concluded that they could simply leave changes in soil carbon out of their life cycle assessments of different livestock management approaches. When they did this it appeared that industrial meat production had a smaller climate impact per unit of meat or milk produced than grass-based production.[11] But the assumption that soil carbon is in equilibrium fails to acknowledge the emissions consequences of management decisions and of land use change. In other words, even if soil carbon will reach equilibrium at a new level 5, 10, or 50 years after a change from cropland to pasture, that does not negate the significance of the carbon that is stored in the soil as a result of that change.[12]

[9] See for example Poeplau and Don (2015).

[10] Dignac et al. (2017), Weil (2017).

[11] For example Capper (2012). One paper compared the greenhouse gas impacts of industrial and grass-based beef both with and without considering soil carbon changes. When they left soil carbon out of the model they found that industrial production did better; when they included soil carbon in the model grass-based production had lower greenhouse gas emissions (Pelletier et al., 2010).

[12] Some authors have suggested that soil carbon gains from changing cropland to pasture would likely cause deforestation for crop production elsewhere, and therefore should not be considered when thinking about how to address livestock's greenhouse gas emissions (Garnett et al., 2017). They do not provide any evidence for such a connection.

Moreover, the notion that soil carbon reaches equilibrium within a few years under a new land use or management is not necessarily valid, for a number of reasons. First, the amount of time it takes to reach that new supposed equilibrium is highly variable depending on soil type and the specific land use and management change, with estimates ranging from 5 years to more than 100 years. Although soil carbon loss following disturbance is often rapid, rebuilding soil carbon usually takes much longer. Second, as mentioned above, most research on soil carbon has focused on the upper 15–30 cm of the soil because that is usually where the concentration of carbon is greatest and because researchers assume that carbon near the soil surface is more likely to be affected by land management. However, when researchers have looked deeper in the soil profile they have found that soil carbon changes significantly both near the surface and down to at least 100 cm, and carbon concentrations at depth can keep changing even after surface carbon concentrations have stabilized.[13] Third, just looking at changes in soil carbon concentrations at a predetermined depth such as up to 15 cm ignores the significance of building or losing surface soil. That carbon-rich A horizon is not exactly 15 cm everywhere. Farmers who pride themselves on building soil health rightly focus on how deep their A horizon is, as well as on the concentration of organic matter[14] in that horizon. If the percentage of organic matter in the A horizon stabilizes at 4% but the depth of the A horizon continues to increase, then total soil carbon is continuing to build. Conversely, if the concentration of organic matter in the A horizon is stable but the depth of the A horizon is slowly decreasing due to erosion, then total soil carbon is likely decreasing.[15]

How does bringing soil carbon into the model affect our understanding of the climate impact of industrial livestock production? When we consider changes in soil carbon, the estimated emissions from industrial livestock go up, especially relative to emissions from grass-fed livestock.[16] Industrial livestock production depends heavily on feeding livestock grain. Grain is high in carbohydrates and low in fiber, so animals can efficiently convert grain feed to the products we want: muscle growth (meat) and milk. But grains are annual crops, which means they die in the autumn and leave the ground without living roots or plant cover between harvest and emergence of the next annual crop. During this interval, which lasts about half the year in much of the U.S., soil is vulnerable to both carbon loss and erosion. In contrast, pasture and range are dominated by perennial plants that cover and hold the soil year-round, and that continue to photosynthesize and move carbon from the atmosphere to the soil later in the fall and earlier in the spring than annual grains. Other

[13] Weil (2017), Sanford et al. (2012).

[14] Farmers usually use the term "organic matter" rather than soil carbon because that is what standard soil analysis measures. Although organic matter analysis doesn't measure all carbon in the soil it provides a good indicator of soil carbon content.

[15] There is speculation that in some cases the carbon in eroded soil may be sequestered when it is deposited in deep water bodies or buried under other sediments, so the carbon fluxes associated with soil erosion are uncertain.

[16] Pelletier et al. (2010), Stanley et al. (2018), Rowntree et al. (2020).

things being equal, soils in perennial diverse grassland contain more carbon than soils in annual crops. And contrary to the assumption that soil carbon is in equilibrium, soils in grain crop rotations continue to lose carbon even after decades of the same management, while soils in healthy grasslands can continue to increase the depth of the carbon-rich upper soil horizon.[17]

5.1.2 Other Greenhouse Gas Emission Differences Among Feed Types

The high-yielding annual grain crops that make up much of the feed for industrial livestock also require more nitrogen fertilizer and pesticides than pasture and range. These inputs are produced using large amounts of fossil fuels, and as discussed in Chap. 3, nitrogen fertilizer applications are associated with emissions of nitrous oxide, a highly potent greenhouse gas.[18]

5.1.3 Overproduction and Waste

While the role of soil carbon and other greenhouse gas emission differences between feed types is important, an even more significant limitation of the efficiency argument is that its basic metric of greenhouse gas emissions per kilogram of product ignores the role of industrial efficiency in driving increased production, consumption, and waste.

The efficiencies of industrial agriculture were not developed to reduce greenhouse gas emissions, rather they are the result of efforts to reduce the cost of production in order to both reduce the cost of meat and dairy and increase income for farmers and food processors. And indeed, the cost of the most efficient meat – chicken – has become remarkably cheap. Overall, the cost of most foods relative to income has declined consistently over the past century.[19] We tend to assume that cheap food will provide greater food security and reduce hunger, both very important goals. However, cheap food has failed to eliminate hunger in the U.S.,[20] and low

[17] Conant et al. (2017), Mogensen et al. (2014), Sanford et al. (2012).

[18] Beste and Idel (2020).

[19] Roser and Ritchie (2021). Their charts also show that the biggest price declines have been for grain, and arguably those price declines and production increases for grain are what is fueling the industrial grain-fed meat system.

[20] The U.S. Department of Agriculture only started tracking food insecurity after 1990, and every year since then at least 10% of American households have experienced food insecurity. During the Great Recession of 2008–2012 nearly 15% of U.S. households reported food insecurity, and although the USDA numbers for 2020 have not yet been released, with the coronavirus pandemic as many as 25% of US households may have suffered food insecurity (Bauer, 2020; Roser & Ritchie 2019).

costs of production have not helped American farms. Instead, cheap, abundant food has led to more food being used, and to more waste.

The tendency for people to consume more of a good when it is produced more efficiently, and therefore more cheaply, is called the Jevons paradox after a nineteenth century British economist who predicted that as coal burning engines became more efficient the use of coal would go up rather than down, because people would use coal power for more purposes.[21] Several scholars have grappled with the question of how this paradox might affect efforts to reduce climate change impacts by increasing energy efficiency, but it has not been used as a critique of the efficiency argument for industrial meat production.[22] Food is often seen as a good with low price elasticity. That is, because we need food to live, people will buy food even when the price is high; and conversely, at least in theory we don't have much reason to buy more food than we need to stay healthy, even if the price is low.

However, it does not follow that because meat is food, demand for livestock products does not depend on price (and efficiency). People choose among different foods based on price all the time. Both critics and supporters of livestock agriculture have noted that as populations become wealthier and meat becomes cheaper, per capita meat consumption rises, because more consumers can afford to eat meat rather than cheaper plant-based proteins. It is also striking that the most efficient (and therefore cheapest) meat type, chicken, has seen the greatest consumption increase both in the U.S. and worldwide with the spread of industrialized poultry production.[23]

In the U.S., where per capita meat consumption was already at 88 kg per year in 1970 what have been the results of 50 years of efficiency gains in livestock production? The answer is complex, but we have certainly not seen agricultural emissions reductions equivalent to the increase in efficiency. First, per capita meat consumption in the US increased to more than 100 kg per year by 2019, a staggering amount, especially considering that that average consumption figure includes children and vegetarians, as well as meat-eating adults.

In fact, fortunately for our health, it is unlikely that Americans are actually eating all that meat. That per capita consumption figure is simply the total amount of meat that "disappeared",[24] divided by the number of people living in the U.S. While most of that meat was eaten by people, some of it was also thrown away, and some was fed to the growing population of pets.

We only have rough estimates of food waste, but there are good reasons to believe that overproduction is driving unhelpful uses of food in a number of ways. The U.S. Department of Agriculture estimates that in the U.S. 10% of food is lost at the

---

[21] Alcott (2005), Freire-González and Puig-Ventosa (2015).

[22] Freire-González and Puig-Ventosa (2015), Alcott (2005).

[23] Winders and Ransom (2019), U.S. Department of Agriculture Economic Research Service (2020).

[24] The U.S. Department of Agriculture tracks meat production, imports, and exports and uses those numbers to calculate what they call disappearance. Disappearance equals production, plus imports, minus exports. Basically, "disappearance" is a crude measure of meat use, but the awkward term acknowledges that we don't know what those uses are.

retail level, and another 21% at the consumer level. This is a very conservative estimate. Significant losses occur before food reaches the retail level that are not included in this statistic.[25] Also, this 31% estimate is based on the assumption that the average American eats 2547 of the 3796 calories available to them per day. This assumption is reasonable for adult males, but for the half of the population that is female this estimate is high, as it is for males under 10 years of age.[26] Other estimates put food waste in North America and Europe as high as 50%.[27]

This food loss has greenhouse gas emission implications at both ends of the food system. On the production end it represents greenhouse gas emissions to produce, process, and transport food that did not actually result in useful nutrition for people. At the other end of the food system methane emissions from landfills account for about a third of total U.S. methane emissions. The primary source of these methane emissions from landfills is food waste. An unknown amount of food is also discarded into the sewage system, likely contributing to more methane emissions.[28]

Food loss is a worldwide problem, but the causes of food loss vary widely between wealthy and less wealthy nations. In less wealthy nations food loss and spoilage are primarily due to lack of good harvesting, storage, and refrigeration capacity. In wealthy nations refrigeration and good storage and transportation infrastructure greatly reduce food spoilage, but the total amount of food loss actually goes up because throughout the food chain more food is discarded due to cosmetic and other issues.

I want to pause here and distinguish between food loss and food waste, and also introduce two additional categories of questionable food use encouraged by efficient industrial food production. As defined by the U.S. Department of Agriculture, **food loss** is the difference between the amount of food available and the amount that is actually eaten by people. So food loss includes unavoidable reductions like loss of moisture in cooking, as well as losses due to spoilage and other discards. **Food waste** is the subset of food loss that is avoidable. Excess consumption – the food people eat that exceeds the amount that is healthy – does not show up as food loss, but arguably it is a form of food waste that has severe public health consequences. Finally, I will talk about the pet food industry.

The first example of food waste that comes to mind is plate waste, the food that makes it all the way to the eater's plate but is discarded because the eater doesn't like the taste or because the serving was too large. The term "waste" implies willful disregard of conservation or at least carelessness, but what food loss is unavoidable or discretionary is not always clear-cut. If I eat chicken, at the end of the meal there are bones on my plate. I might simmer those bones to make broth, but still I will

---

[25] Bloom (2010) and Minor (2020).

[26] https://health.gov/our-work/food-nutrition/2015-2020-dietary-guidelines/guidelines/appendix-2/#table-a2-1 U.S. Department of Health Dietary Guidelines.

[27] Stuart (2009) and Papargyropoulou et al. (2014).

[28] Buzby, Farah-Wells, and Hyman (2014), Committee on a Systems Approach to Reducing Consumer Food Waste et al. (2020).

eventually send the bones to the landfill. I consider the bones to be inedible, so to my mind they are food loss, but not waste. But in rural South Africa and probably other parts of the world people do eat chicken bones, and discarding them would be considered waste. And indeed, the bones contain valuable nutrients.

Chicken bones are only one of many examples where people might disagree on whether a food is suitable for human consumption. In our food system there is a lot of confusion around when a food is no longer good to eat. Most processed foods are labeled with a date, but the meaning of that date varies. "Use by" and "best if used by" dates are intended to let the consumer know that the food is at peak quality at least until that date. "Sell by" dates are intended to help retailers ensure that the foods they stock will maintain peak quality for at least a few days after they are sold. Consumers often think that the foods are unsafe to eat after the "sell by" or "best by" dates and so discard those foods. But in fact in most cases the foods are still wholesome when those dates have passed. Retailers likewise often err on the side of caution and throw away foods before the sell-by date. Many retailers also refuse to donate "expired" foods to food pantries, even though legislation protects businesses from liability for illness from donated foods.[29] There are legitimate concerns and uncertainties about food safety that all steps in the food chain have to navigate, and when meat and dairy are plentiful and cheap it appears economically rational at every step to buy more than is needed and discard anything of questionable value.

Discounting products sold in large quantities is another artifact of efficient overproduction that leads to waste. The price for a half gallon (roughly 2 liters) of milk in a nearby grocery store is $1.89; the price for a whole gallon is $2.59. With those prices the customer saves money by buying a full gallon, even if they wind up throwing away a quart (1 liter) when the milk goes sour before they drink it all. We have similar bulk pricing and incentives to waste across many foods in the U.S. The cost of packaging accounts for only part of the price differential; the rest of the price difference can only be explained by a desire to get consumers to buy more than they really need.

At a broader level, some argue that feeding grains and other foods that could be eaten by humans to livestock is in itself a form of planned food waste, because even that most efficient animal, the Cornish Cross chicken, uses nearly half of the nutrients it consumes for its own metabolic needs.[30] Certainly the increased efficiency of modern grain production encourages the practice of feeding livestock a grain-heavy diet, which then creates more demand and drives yet more grain production. Because much of the relative efficiency of intensive livestock production versus grass-based meat comes from feeding edible grains, this viewpoint that feeding grain to livestock is a form of waste offers a fundamental challenge to the overall efficiency of industrial livestock. But even setting aside that fundamental challenge, there is

---

[29] Leib et al. (2013), Committee on a Systems Approach to Reducing Consumer Food Waste et al. (2020), Bloom (2010), Gunders (2012).

[30] Papargyropoulou et al. (2014).

another major way that efficient and cheap meat promotes increased consumption that is rarely discussed.

When I was a child in the 1960s and 1970s we regularly ate parts of the animal that rarely make it to American dinner tables today, including liver, tongue, and heart. They were not the preferred cuts, either because they had strong flavors or because they took more preparation than other cuts, so they were relatively cheap. But as all meat became cheap, those nutritious parts of the animal stopped showing up on American dinner tables. Today you cannot find those cuts, much less kidneys or tripe, in most American grocery stores. Instead, they are used to make food for the growing population of cats and dogs.[31] Pets now eat about a third of the meat produced in the U.S., and increasingly people are buying premium grain-free dog and cat foods that contain even more meat than standard pet foods.[32] In addition to accounting for a significant portion of the meat use in the U.S., pets also produce feces that presumably contribute to the methane emissions from landfills. In this context it is curious that, while there are many popular and scholarly articles calling for people to reduce or end meat consumption for the health of the planet, there is almost no discussion of reducing pet populations for environmental reasons.

It is likely that several factors contribute to the growth of the pet population in the U.S., including a growth in the number of households as the average number of people per household shrinks. But the availability of convenient, cheap pet food is probably also a factor. While the 33% of American meat that cats and dogs eat isn't wasted in the traditional sense of the word, it also isn't feeding humans. And we do not have good statistics on how much pet food is wasted, though we do know that increasing numbers of cats and dogs are overweight.[33]

5.1.3.1 Excess Consumption

Even with the amount of meat thrown away and diverted to pets, Americans eat more meat than is recommended.[34] More than 40% of U.S. adults suffer from obesity, and the numbers keep going up.[35] Obesity is also an increasing health issue around the world. While simple carbohydrates are major contributors to excess

---

[31] From 1988 to 2016 the number of pet dogs in the U.S. increased by around 50% and the number of cats increased by nearly 80% (Rowan, 2018).

[32] Okin (2017). While some of the meat that goes to pet food is not edible by humans without processing, much of it was valued before the era of cheap meat.

[33] Verbrugghe (2018).

[34] U.S. Department of Agriculture, & U.S. Department of Health and Human Services (2020). Moreover, Nestle and others have noted that U.S. nutritional guidelines are based on lobbying by farm and food industry groups as well as on nutritional science, and that as a result the recommendations for meat and dairy consumption are higher than nutrition science alone might indicate (Nestle, 2007).

[35] Hales (2020).

weight gain, excess meat consumption may also play a role. Health professionals and nutritionists are still trying to understand the complex factors contributing to obesity, but there is a growing recognition that food marketing is surprisingly successful at overcoming the body's ability to signal satiety and adequate nutrition, and commodity groups and food retailers work hard to persuade people to buy more meat and dairy.

5.1.4 *The Need for Better Policy*

Between erring on the side of caution, competitive marketing that celebrates abundance, plate waste, overconsumption, and pet food, efficient and therefore cheap meat production encourages overuse and outright waste in the U.S. And we keep increasing production, even though on average Americans already consume staggering amounts of meat. Certainly, the U.S. lacks a good mechanism for stabilizing or reducing production when healthy demand is met. The classical "law" of supply and demand is constrained by several factors, including federal farm policy, farm culture, and biology.

In most cases when farm prices go down and farmers lose income as a result, their response is not to decrease production. Rather they seek to maintain their income by producing even more, or by finding ways to reduce their production costs while maintaining production levels. U.S. farm policy supports farmers' approach of maintaining high production levels through government-subsidized insurance programs such as Dairy Margin Coverage and Price Loss Coverage that hedge against low prices. In addition, the federal government facilitates huge advertising campaigns to get people to eat more livestock products than they need in a frantic attempt to keep demand up and so keep prices above the cost of efficient production for most farmers. For every steer or cow they sell for meat, American farmers must pay $1 to the Beef Checkoff program, a marketing program "designed to increase the demand for beef at home and abroad."[36] There are similar mandatory checkoffs to support marketing programs for pork, eggs, and dairy.[37]

Farm culture celebrates high production in many ways, including annual grain yield contests and livestock competitions that favor growth and production traits over traits associated with animal health and welfare. Many American farm groups continue to simultaneously promote the contradictory notions that farmers' high production is needed to feed the world, and that low meat and dairy prices are the fault of misguided consumers who choose plant-based beverages and proteins over

---

[36] U.S. Department of Agriculture Agricultural Marketing Service (n.d.).

[37] These checkoff programs also actively market themselves to the farmers that pay for them (U.S. Department of Agriculture Agricultural Marketing Service n.d.; Beef Checkoff 2021; Sanburg 2021). https://www.media/press-releases

milk and meat rather than a result of overproduction.[38] At the same time Farm Bureau proudly announces on its website that "Compared to 1990, U.S. milk production has increased by 71%, beef production has increased by nearly 50% and pork production has increased by 17%."[39] During that same period the U.S. population increased by 24%, but there is no mention of the disparity between these production increases and the smaller population growth or of the role that disparity might play in keeping farm prices low.

Throughout this period, Farm Bureau and U.S. dairy groups' fierce criticism of Canadian trade policies offers a striking illustration of the reality that feeding the hungry is not the true purpose of the dairy industry's push for more production.[40] Canada has no shortage of dairy products, but because of its supply management policies designed to protect Canadian farmers, Canada's wholesale dairy prices are higher than those in the U.S. American farmers want access to those higher prices; but if they get that access without managing their own production levels they will almost certainly drive those prices down to U.S. levels.

And biology also hampers the ability of many farmers to adjust production to demand. It takes 1–2 years for a beef animal to reach slaughter size. Likewise, it takes around 2 years for a dairy cow to be ready to milk. When prices are low, beef and dairy farmers cannot temporarily stop feeding and milking their animals. They could breed and raise fewer replacement animals, but they hope prices will be better in a year or two, and so the cycle of overproduction continues.

This current system of constantly increasing livestock production isn't working very well, either for the climate or for farmers.[41] In my home state of Wisconsin, which proudly calls itself the dairy state, the last 5 years have been devastating for dairy farmers. In 2019 alone Wisconsin lost 10% of its dairy farms after years of depressed milk prices.[42] Some went bankrupt; others were able to cover their debts by selling their cows and/or their land. American dairy farmers and the bankers they work with have become accustomed to bad years when the price they get paid for their milk falls below their cost of production, but even with government subsidies and understanding bankers, when low prices go on for years some farms cannot keep going. In the past 26 years there were only 2 years when the farm gate price of

[38] Farm Bureau offers a special "Women's Communications Bootcamp" to teach farm women how to promote positive messages about American agriculture, as well as a wide array of resources aimed at schoolchildren and consumers (Farm Bureau n.d.). The most receptive audience for these messages is probably their own membership. Other farm groups have similar programs (Wisconsin Milk Marketing Board 2021). USDA looked at the relationship between consumption of plant-based "milks" and found that while their use is growing, that growth only accounts for about one fifth of the decline in per capita fluid milk consumption (Stewart et al., 2020).

[39] Interestingly, they do not include emissions from feed production in their discussion of livestock greenhouse gas emissions, only enteric and manure emissions (Newton 2021).

[40] Laca (2020). https://www.dairyherd.com/news/exports/dairy-leaders-praise-ustr-action-canadian-quota-enforcement

[41] MacDonald et al. (2020).

[42] Kottke (2020).

milk was higher than the cost of production for the average dairy farm.[43] In our current system, though, the failure of those farms doesn't lead to a reduction or even a leveling off of the milk supply. In 2019, as dairy farms across the country were going bankrupt, U.S. milk production was 0.4% higher than in 2018. And despite supply chain disruptions and reports of milk being dumped early in the pandemic, milk production in 2020 was about 2% higher than in 2019.[44] The best cows from the farms that go out of business are bought by their better financed neighbors, and the relentless increase in milk production continues.

The American Farm Bureau and Farmers Union are two national farm groups that go back more than a century. The Farm Bureau tends to be politically conservative in its policy positions, and has consistently celebrated the large production gains of industrial agriculture. Farmers Union from the start has been more inclined to challenge the idea that unfettered capitalism is always good for farmers, and a few years ago the Wisconsin chapter of Farmers Union asked economists at the University of Wisconsin-Madison and Cornell University to explore how supply management policies for dairy might affect the farm economy and consumers. Their models of two different supply management structures found that both would have improved prices for farmers and prevented farm bankruptcies, with slight impacts on consumer prices and a modest decrease in production.[45] Although National Farm Bureau remains skeptical about supply management, the Wisconsin Farm Bureau chapter and the mainstream agricultural press have taken note of the Farmers Union "Dairy Together" campaign and are cautiously interested.[46]

Supply management can potentially have negative impacts, if it is not carefully designed. It could drive up food prices, and it could further consolidate the position of large farmers and prevent new farmers from starting. However, it is important to recall that our current agricultural system is already not a free market. Our overproduction of grain and livestock products is perpetuated by government subsidies that have averaged nearly $17 billion per year over the last 25 years and topped $22 billion in 2019.[47] To the frustration of many grassroots farmers and activists these subsidies overwhelmingly support the largest farms and leave out vegetable and fruit farms and non-traditional farming approaches. Despite our excess agricultural production, more than 10% of American households experience food insecurity even in good years, and during the economic downturns of 2008–2012 and 2020 as many as 15 and 25% of households reported being unable to consistently afford the food they need without sacrificing other basic needs like shelter, utilities, or medical

[43] MacDonald et al. (2020).

[44] U.S. Department of Agriculture, National Agricultural Statistics Service (2020, 2021) https://www.gov/Statistics_by_State/Wisconsin/Publications/WIMilkCty19.pdf

[45] Nicholson and Stephenson (2019). The report also found that both supply management approaches would have reduced government subsidy expenditures by $329 million to $1245 million over the 5 year period modeled.

[46] Laca (2018), Schleis (2020), Wisconsin Farmers Union (2020).

[47] Environmental Working Group (n.d.).

care.[48] The U.S. government already influences supply of milk, meat, and grain, just not in a way that benefits most farmers or the environment. It is time for a re-examination of the impacts of farm policy on production, farm viability, and the climate.

5.1.5 Variability and the Special Case of Beef

Most comparisons between efficient industrial agriculture and extensive grass-based systems come up with a single number for each system. For example, Capper reported in her widely cited paper that the greenhouse gas emissions for grassfed beef were about 1.7 times greater than emissions for conventional, intensive beef. This type of comparison hides the wide range of greenhouse gas emissions, as well as other environmental impacts, that exist within both the efficient, industrial and the grass-based livestock production systems.[49] Some of the variation in emissions is the result of local climate and soils, but much of it also comes from specifics of manure and crop management, practices over which individual farmers have some control.[50] In most cases it is more feasible for a farmer to make improvements within the production system they have than to switch to a completely different approach. So our public discourse about agriculture and climate needs to devote more attention to what farmers can do to reduce their greenhouse gas emissions *within* either the industrial or grass-fed approach, and spend less time arguing about which system is better on average. Chapter 8 will try to outline some suggestions for that approach.

But there is one area of significant variation in greenhouse gas emissions within the efficient industrial paradigm that poses a special challenge for the ongoing debate about whether industrial or grass-based meat is better. Poultry and pork are far more efficient at converting their feed to meat than ruminants. It takes about 2 kg of feed to produce a kg of chicken, about 3 kg of feed to produce a kg of pork, and more than 6 kg of feed to produce a kg of beef.[51] Because pigs and chickens are not ruminants, they also emit very little methane as part of their digestive process. As a result, the greenhouse gas emission intensity for industrial poultry in North America is around 4.5 kg CO_2eq/kg meat, for pork is around 5.5, and for beef is more than

[48] Bauer (2020).

[49] Aguirre-Villegas, Larson, and Reinemann (2015), Wattiaux et al. (2019).

[50] Capper (2012), Rotz et al. (2019). Rotz's paper shows a wide range of emissions and water impacts for beef production for each region. He identifies enteric emissions as the key factor affecting greenhouse gas emissions, with local climate as another important variable, but says each farm's situation is different, so concludes that "Improvements must be made on an individual operation basis."

[51] Because cattle diets are highly variable across the animal's life, estimates of cattle feed efficiency are also variable, ranging from a factor of 6 to more than 10.

30.[52] (This figure is for beef from animals raised for meat; beef from dairy cows has an emission intensity of around 10 kg CO_2eq/kg meat.) Moreover, while cattle gestation lasts 9 months and beef cattle typically take 1–2 years to grow from birth to slaughter weight, chicken eggs incubate for about 3 weeks, and the chicks go from hatching to slaughter weight in 2–3 months. This fast poultry reproduction and growth rate make it easier to adjust supply in response to fluctuations in meat demand or grain availability.

So going back to my neighbor's question about which meat is least harmful from a climate perspective, as seen through the efficiency lens the answer is clearly poultry, closely followed by pork. Although industrial beef finished in a feedlot performs better than grass-finished beef from the efficiency viewpoint, its greenhouse gas emissions are still far higher than those of poultry and pork. Beef advocates point out that, unlike poultry and pigs, over their life even feedlot ruminants typically get roughly 80% of their feed from forages that humans cannot eat, so it is not fair to compare their feed efficiency to that of non-ruminant animals that primarily eat foods that could also feed humans directly. However, with this reasoning you cannot also conclude that feedlot beef is better for the climate than grass-finished beef, even though it is more efficient. During the feedlot finishing stage more than half of the cattle's feed is grain. Another 30% is by-products, most of which chicken and pigs can eat, and some of which could probably be processed into human food.[53]

5.1.6 The Special Case of Dairy

When we look at ruminants, however, dairy animals provide a special case. While to a large extent, eaters have been willing to switch from beef to poultry, ruminants are the only farm animals that produce commercial quantities of milk, and so far, most consumers have not been willing to switch to plant-based cheese and milk substitutes. In fact, after reports in the 1990s that trans fats in margarine pose similar or greater heart disease risks as butterfat, per capita U.S. margarine consumption declined precipitously and butter consumption increased slightly.[54] The emission intensities for dairy products are also considerably lower than those for beef, with milk in North America emitting less than 2 kg CO_2eq per kg of milk, compared to an emission intensity for North American beef of more than 30 kg CO_2eq per kg of meat. Of course, milk contains a lot of water, but even when measured per kilo of protein in milk or meat, the emissions associated with meat are nearly five times higher than those for milk.[55]

[52] MacLeod et al. (2013), Opio et al. (2013). Worldwide, the emission intensities for pork and poultry are a little over 5 kg CO_2eq/ kg meat for poultry and 6 kg CO_2eq/ kg meat for pork; for beef they exceed 60 kg CO_2eq/ kg meat.

[53] National Academies of Sciences, Engineering, and Medicine (2016).

[54] Stewart et al. (2020).

[55] Opio et al. (2013).

Beef from dairy cattle also has a lower carbon hoofprint than beef from cattle that are not milked. Both in North America and globally the emissions intensity for meat from dairy cattle is roughly one third that of beef herds.[56] That dairy beef comes from cull cows, including cows that have trouble conceiving,[57] and cows that the farmer does not want to keep for a variety of other reasons, from health or behavior problems to low milk production. Dairy beef also comes from male calves and those female calves that are not needed to maintain the milking herd. For all these reasons, the question about whether industrial or grass-based systems are more climate-friendly makes more sense for dairy farms than for pure beef operations.

While many studies conclude that industrial dairies where the cows are kept indoors and fed carefully formulated rations have a smaller climate impact than grass-based dairies, mainly because each cow produces more milk,[58] other studies conclude that blended systems have the optimum climate performance, even without including soil carbon in the analysis. In these blended systems cows graze when the grass is growing well, but they also receive grain supplements when they are milked. The grain supplements help maintain high milk yield and reduce enteric methane emissions, while producing the feed the cows get from pasture typically has lower greenhouse gas emissions than raising harvested forages.[59]

One potential climate advantage that large confined dairies have is that installing and operating a manure digester is more likely to be economically viable than for small or grass-based farms. Manure digesters reduce greenhouse gas emissions both by preventing the manure methane from escaping to the atmosphere, and also by generating electricity, transportation fuel, or heat using the captured methane, which reduces the demand for electricity or heat produced from fossil-fuels.[60] When cows are on pasture their manure does not emit methane, but it also does not provide electricity or heat. However, solar electricity generation is now more cost-effective than manure digestion, and solar panels can be installed regardless of farm type or size.

5.2 Closing Thoughts

Efficiency is an important part of animal agriculture's response to the challenge of climate change, but unless there is an equitable and effective way to stop production when we have enough meat and milk, efficiency will not reduce livestock's greenhouse gas emissions.

---

[56] Opio et al. (2013).

[57] Milk production in cows peaks about 2 months after giving birth and then declines steadily, so to maintain milk production dairy cows need to have a calf every year.

[58] Dutreuil et al. (2014), Kiefer et al. (2014), Capper et al. (2009).

[59] Rotz et al. (2020), Aguirre-Villegas et al. (2017).

[60] Aguirre-Villegas et al. (2015, b, 2017).

While this book and this chapter focus on the greenhouse gas emissions associated with animal agriculture, we must also keep in mind the other environmental and social consequences of our agricultural approaches. Efficient industrial livestock production has well-documented consequences for worker health, animal welfare, water quality, and local air quality.[61] This is not to say that all large concentrated animal feeding operations (CAFOs) are bad or that all small grazing farms are good. Some large industrial farms do a good job of minimizing pollution; provide better pay, benefits, and working conditions to their employees than most small farms can afford; and strive to keep their animals healthy and comfortable. Some small extensive farms cause pollution, fail to provide decent pay and good working conditions to the farm owners or their employees, and have animal welfare problems because they lack the economic capacity or skill to manage all those challenges. But where we see the high concentrations of livestock associated with efficient industrial agriculture we very often also see problems, such as widespread surface and drinking water contamination, reduced air quality, habitat loss, and public health impacts.[62] Efficient industrial systems also require capital investment, creating barriers for less wealthy farmers. These barriers exist even within wealthy nations with agricultural subsidy programs, and they are likely to be insurmountable for the majority of farmers in much of Asia, Africa, and South America.[63]

The COVID-19 pandemic has also illustrated tensions between efficiency, justice, and resilience in the supply chain for industrial meat. Industrial slaughterhouses designed for maximum efficiency and speed of processing, with employees working shoulder-to-shoulder, also turned out to be highly efficient at spreading disease, and although U.S. meatpackers were encouraged to adopt policies to reduce disease transmission, they were also exempted from liability for worker illness.[64] Efficient food processing specialization also meant that when food service and restaurants closed and their demand fell, farmers whose products normally went to food service were forced to dump milk, even though grocery stores were unable to keep up with increased individual consumer demand for dairy products.[65]

As we work with farmers and agribusiness to reduce our carbon hoofprint, we also need to include water, air, and soil quality; worker health and dignity; and animal welfare in our core requirements for all agriculture, rather than treating them as afterthoughts or externalities.

[61] Ashwood et al. (2014), Burkholder et al. (2007), Carrel et al. (2016), Greger and Koneswaran (2010).

[62] There is an enormous literature on the negative environmental and social impacts of many concentrated animal feeding operations, including the above and Waller et al. (2021).

[63] Shiller (2017), Kriegl and McNair (2005), Carlisle et al. (2019).

[64] Hendrickson (2020), Taylor et al. (2020).

[65] Hendrickson (2020), Orden (2020), Garcés (2020).

Bibliography

Aguirre-Villegas, H. A., Larson, R., & Reinemann, D. J. (2015). Effects of management and co-digestion on life cycle emissions and energy from anaerobic digestion: Modeling and analysis: Effects of management and co-digestion on life cycle emissions and energy from anaerobic digestion. *Greenhouse Gases: Science and Technology, 5*(5), 603–621. https://doi.org/10.1002/ghg.1506

Aguirre-Villegas, H. A., Passos-Fonseca, T. H., Reinemann, D. J., Armentano, L. E., Wattiaux, M. A., Cabrera, V. E., Norman, J. M., & Larson, R. (2015). Green cheese: Partial life cycle assessment of greenhouse gas emissions and energy intensity of integrated dairy production and bioenergy systems. *Journal of Dairy Science, 98*(3), 1571–1592. https://doi.org/10.3168/jds.2014-8850

Aguirre-Villegas, H. A., Passos-Fonseca, T. H., Reinemann, D. J., & Larson, R. (2017). Grazing intensity affects the environmental impact of dairy systems. *Journal of Dairy Science, 100*(8), 6804–6821. https://doi.org/10.3168/jds.2016-12325

Alcott, B. (2005). Jevons' paradox. *Ecological Economics, 54*, 9–21. https://doi.org/10.1016/j.ecolecon.2005.03.020

Ashwood, L., Diamond, D., & Thu, K. (2014). Where's the farmer? Limiting liability in midwestern industrial hog production. *Rural Sociology, 79*(1), 2–27. https://doi.org/10.1111/ruso.12026

Bauer, L. (2020). *Hungry at thanksgiving: A fall 2020 update on food insecurity in the US* [Blog post]. Brookings Institute. https://www.brookings.edu/blog/up-front/2020/11/23/hungry-at-thanksgiving-a-fall-2020-update-on-food-insecurity-in-the-u-s/

Beef Checkoff. (2021, July 22). Nutrition research improves public perception of beef. *Morning AgClips*. https://www.morningagclips.com/nutrition-research-improves-public-perception-of-beef/

Beste, A., & Idel, A. (2020, June 1). The myth of climate smart agriculture – Why Less Bad Isn't Good. *ARC* 2020. https://www.arc2020.eu/the-myth-of-climate-smart-agriculture-why-less-bad-isnt-good/

Bloom, J. (2010). *American wasteland: How America throws away nearly half of its food (and what we can do about it)*. Da Capo Lifelong Books.

Burkholder, J., Libra, B., Weyer, P., Heathcote, S., Kolpin, D., Thorne, P. S., & Wichman, M. (2007). Impacts of waste from concentrated animal feeding operations on water quality. *Environmental Health Perspectives, 115*(2), 308–312. https://doi.org/10.1289/ehp.8839

Buzby, J.C., Farah-Wells, H., & Hyman, J. (2014). The estimated amount, value, and calories of postharvest food losses at the retail and consumer levels in the United States. *USDA-ERS Economic Information Bulletin*, (121).

Capper, J. L. (2012). Is the grass always greener? Comparing the environmental impact of conventional, natural and grass-fed beef production systems. *Animals, 2*(4), 127–143. https://doi.org/10.3390/ani2020127

Capper, J. L., Cady, R. A., & Bauman, D. E. (2009). The environmental impact of dairy production: 1944 compared with 2007. *Journal of Animal Science, 87*(6), 2160–2167. https://doi.org/10.2527/jas.2009-1781

Carlisle, L., de Wit, M. M., DeLonge, M. S., Calo, A., Getz, C., Ory, J., Munden-Dixon, K., Galt, R., Melone, B., Knox, R., Iles, A., & Press, D. (2019). Securing the future of US agriculture: The case for investing in new entry sustainable farmers. *Elementa: Science of the Anthropocene, 7*, 17. https://doi.org/10.1525/elementa.356

Carrel, M., Young, S., & Tate, E. (2016). Pigs in space: Determining the environmental justice landscape of swine concentrated animal feeding operations (CAFOs) in Iowa. *International Journal of Environmental Research and Public Health, 13*(9), 849. https://doi.org/10.3390/ijerph13090849

Committee on a Systems Approach to Reducing Consumer Food Waste, Board on Environmental Change and Society, Food and Nutrition Board, Division of Behavioral and Social Sciences and Education, Health and Medicine Division, & National Academies of Sciences,

Engineering, and Medicine. (2020). *A national strategy to reduce food waste at the consumer level* (B. O. Schneeman & M. Oria, Eds.; p. 25876). National Academies Press. https://doi.org/10.17226/25876.

Conant, R. T., Cerri, C. E., Osborne, B. B., & Paustian, K. (2017). Grassland management impacts on soil carbon stocks: A new synthesis. *Ecological Applications, 27*(2), 662–668.

Dairy Business Association and Edge Dairy Farmer Cooperative. (2020, November 10). Dairy groups push back on anti-CAFO resolution. *Morning AgClips*. https://www.morningagclips.com/dairy-groups-push-back-on-anti-cafo-resolution/

Dignac, M.-F., Derrien, D., Barré, P., Barot, S., Cécillon, L., Chenu, C., Chevallier, T., Freschet, G. T., Garnier, P., Guenet, B., Hedde, M., Klumpp, K., Lashermes, G., Maron, P.-A., Nunan, N., Roumet, C., & Basile-Doelsch, I. (2017). Increasing soil carbon storage: Mechanisms, effects of agricultural practices and proxies. A review. *Agronomy for Sustainable Development, 37*(2), 14. https://doi.org/10.1007/s13593-017-0421-2

Dutreuil, M., Wattiaux, M., Hardie, C. A., & Cabrera, V. E. (2014). Feeding strategies and manure management for cost-effective mitigation of greenhouse gas emissions from dairy farms in Wisconsin. *Journal of Dairy Science, 97*(9), 5904–5917. https://doi.org/10.3168/jds.2014-8082

Environmental Working Group. (n.d.). *Total USDA – Subsidies*. Retrieved December 11, 2020, from https://farm.ewg.org/progdetail.php?fips=00000&progcode=total&page=states

Farm Bureau. (n.d.). *Women's Communications Boot Camp*. Farm Bureau. Retrieved December 4, 2021, from https://www.fb.org/programs/womens-leadership-program/communications-boot-camp/

Freire-González, J., & Puig-Ventosa, I. (2015). *Energy efficiency policies and the Jevons paradox*. 5(1), 12.

Garcés, L. (2020). *COVID-19 exposes animal agriculture's vulnerability*. Agriculture and Human Values. https://doi.org/10.1007/s10460-020-10099-5.

Garnett, T., Godde, C., Muller, A., Röös, E., Smith, P., & de Boer, I. (2017). *Ruminating on cattle, grazing systems, methane, nitrous oxide, the soil carbon sequestration question – and what it all means for greenhouse gas emissions*, p. 127.

Greger, M., & Koneswaran, G. (2010). The public health impacts of concentrated animal feeding operations on local communities. *Family & Community Health, 33*(1), 11–20. https://doi.org/10.1097/FCH.0b013e3181c4e22a

Gunders, D. (2012). *Wasted: How America is losing up to 40 percent of its food from farm to fork to landfill* (p. 26). Natural Resources Defense Council. https://www.nrdc.org/sites/default/files/wasted-food-IP.pdf

Hales, C. M. (2020). *Prevalence of obesity and severe obesity among adults: United States, 2017–2018*. 360, 8.

Hendrickson, M. K. (2020). Covid lays bare the brittleness of a concentrated and consolidated food system. *Agriculture and Human Values, 37*(3), 579–580. https://doi.org/10.1007/s10460-020-10092-y

Kiefer, L., Menzel, F., & Bahrs, E. (2014). The effect of feed demand on greenhouse gas emissions and farm profitability for organic and conventional dairy farms. *Journal of Dairy Science, 97*(12), 7564–7574. https://doi.org/10.3168/jds.2014-8284

Kottke, C. (2020, January 31). *Wisconsin continues to lead U.S. in farm bankruptcies*. Wisconsin State Farmer. https://www.wisfarmer.com/story/news/2020/01/31/wisconsin-continues-lead-u-s-farm-bankruptcies/4625866002/

Kriegl, T., & McNair, R. (2005). *Pastures of plenty. Center for integrated agricultural systems*. University of Wisconsin-Madison. https://www.cias.wisc.edu/wp-content/uploads/2008/07/pastplenty607.pdf

Laca, A.-L. (2018, December 4). *Wisconsin Farm Bureau votes to consider dairy supply management*. Dairy Herd Management. https://www.dairyherd.com/news/wisconsin-farm-bureau-votes-consider-dairy-supply-management

Laca, A.-L. (2020, December 9). *Dairy leaders praise USTR action on Canadian quota enforcement*. Dairy Herd Management. https://www.dairyherd.com/news/exports/dairy-leaders-praise-ustr-action-canadian-quota-enforcement

Lal, R. (2002). Soil carbon dynamics in cropland and rangeland. *Environmental Pollution, 10*.

Leib, E. B., Gunders, D., Ferro, J., Nielsen, A., Nosek, G., & Qu, J. (2013). *The dating game: How confusing food date labels lead to food waste in America*. Natural Resources Defense Council and Harvard Food Law and Policy Clinic. https://www.nrdc.org/sites/default/files/dating-game-report.pdf

MacDonald, J. M., Law, J., & Mosheim, R. (2020). *Consolidation in U.S. dairy farming*, p. 61.

MacLeod, M., Gerber, P., Mottet, A., Tempio, G., Falcucci, A., Opio, C., Vellinga, T., Henderson, B., & Steinfeld, H. (2013). *Greenhouse gas emissions from pig and chicken supply chains: A global life cycle assessment*. Food and Agriculture Organization of the United Nations. http://www.fao.org/3/i3460e/i3460e00.htm

Magdoff, F., & van Es, H. (2021). *Building soils for better crops ecological management for healthy soils* (4th ed.). USDA SARE. https://www.sare.org/resources/building-soils-for-better-crops/

Minor, T. (2020). *Economic Drivers of Food Loss at the Farm and Pre-Retail Sectors: A Look at the Produce Supply Chain in the United States* (Economic Information Bulletin No. (EIB-216); p. 39). U.S. Department of Agriculture, Economic Research Service. https://www.ers.usda.gov/publications/pub-details/?pubid=95778.

Mogensen, L., Kristensen, T., Nguyen, T. L. T., Knudsen, M. T., & Hermansen, J. E. (2014). Method for calculating carbon footprint of cattle feeds – Including contribution from soil carbon changes and use of cattle manure. *Journal of Cleaner Production, 73*, 40–51. https://doi.org/10.1016/j.jclepro.2014.02.023

National Academies of Sciences, Engineering, and Medicine. (2016). *Nutrient requirements of beef cattle* (8th ed.). National Academies Press.

Nestle, M. (2007). *Food politics: How the food industry influences nutrition and health*. University of California Press.

Newton, J. (2021, February 22). Previewing 2019 Agricultural Emissions. *Farm Bureau*. https://www.fb.org/market-intel/previewing-2019-agricultural-emissions

Nicholson, C., & Stephenson, M. (2019). *Analyses of selected dairy programs proposed to reduce variability in milk prices and farm income* (DMAP Working Paper 19-01, p. 44). https://dairy-markets.org/PubPod/Podcast/Misc/SupplyManagement/DMAP%20Working%20Paper%2019-01.pdf

Okin, G. S. (2017). Environmental impacts of food consumption by dogs and cats. *PLoS One, 12*(8), p.e0181301.

Opio, C., Gerber, P., Mottet, A., Falcucci, A., Tempio, G., MacLeod, M., Vellinga, T., Henderson, B., & Steinfeld, H. (2013). Greenhouse gas emission from ruminant supply chains: A global life cycle assessment. *Food and Agriculture Organization of the United Nations*. http://www.fao.org/3/i3461e/i3461e00.htm

Orden, D. (2020). Resilience and vulnerabilities of the North American Food system during the Covid-19 pandemic. EuroChoices, 1746-692X.12273. https://doi.org/10.1111/1746-692X.12273.

Papargyropoulou, E., Lozano, R., Steinberger, K., Wright, N., & Ujang, Z. B. (2014). The food waste hierarchy as a framework for the management of food surplus and food waste. *Journal of Cleaner Production, 76*, 106–115. https://doi.org/10.1016/j.jclepro.2014.04.020

Pelletier, N., Pirog, R., & Rasmussen, R. (2010). Comparative life cycle environmental impacts of three beef production strategies in the Upper Midwestern United States. *Agricultural Systems, 103*(6), 380–389. https://doi.org/10.1016/j.agsy.2010.03.009

Poeplau, C., & Don, A. (2015). Carbon sequestration in agricultural soils via cultivation of cover crops – A meta-analysis. *Agriculture, Ecosystems & Environment, 200*, 33–41. https://doi.org/10.1016/j.agee.2014.10.024

Roser, M., & Ritchie, H. (2019). Hunger and Undernourishment. *Our World in Data*. https://ourworldindata.org/hunger-and-undernourishment

Roser, M., & Ritchie, H. (2021). Food Prices. *Our World in Data*. https://ourworldindata.org/food-prices

Rotz, C. A., Asem-Hiablie, S., Place, S., & Thoma, G. (2019). Environmental footprints of beef cattle production in the United States. *Agricultural Systems, 169*, 1–13. https://doi.org/10.1016/j.agsy.2018.11.005

Rotz, C. A., Holly, M., de Long, A., Egan, F., & Kleinman, P. J. A. (2020). An environmental assessment of grass-based dairy production in the northeastern United States. *Agricultural Systems, 184*, 102887. https://doi.org/10.1016/j.agsy.2020.102887

Rowan, A. N. (2018). *Companion animal statistics in the USA*. Demography and Statistics for Companion Animal Populations Collection, 7, 15.

Rowntree, J. E., Stanley, P. L., Maciel, I. C. F., Thorbecke, M., Rosenzweig, S. T., Hancock, D. W., Guzman, A., & Raven, M. R. (2020). Ecosystem Impacts and Productive Capacity of a Multi-Species Pastured Livestock System. *Frontiers in Sustainable Food Systems, 4*, 13.

Sanburg, H. (2021, February 23). A World Without The Beef Checkoff. *Drovers*. https://www.drovers.com/news/industry/sanburg-world-without-beef-checkoff

Sanderman, J., Hengl, T., & Fiske, G. J. (2018). Correction for Sanderman et al., Soil carbon debt of 12,000 years of human land use. *Proceedings of the National Academy of Sciences, 115*(7), E1700. https://doi.org/10.1073/pnas.1800925115

Sanford, G. R., Posner, J. L., Jackson, R. D., Kucharik, C. J., Hedtcke, J. L., & Lin, T.-L. (2012). Soil carbon lost from Mollisols of the North Central U.S.A. with 20 years of agricultural best management practices. *Agriculture, Ecosystems & Environment, 162*, 68–76. https://doi.org/10.1016/j.agee.2012.08.011

Schleis, J. (2020, November 25). The case for supply management. In *Central Time*. Wisconsin Public Radio. https://www.wpr.org/case-supply-management-wisconsin-dairy

Shiller, B. (2017, August 30). What does it cost to start a new farm? *Fast Company*. https://www.fastcompany.com/40458330/what-does-it-cost-to-start-a-new-farm

Stanley, P. L., Rowntree, J. E., Beede, D. K., DeLonge, M. S., & Hamm, M. W. (2018). Impacts of soil carbon sequestration on life cycle greenhouse gas emissions in Midwestern USA beef finishing systems. *Agricultural Systems, 162*, 249–258. https://doi.org/10.1016/j.agsy.2018.02.003

Steinfeld, H., Gerber, P., Wassenaar, T. D., Castel, V., Rosales M., M., & Haan, C. de. (2006). *Livestock's long shadow: Environmental issues and options*. Food and Agriculture Organization of the United Nations.

Stewart, H., Kuchler, F., Cessna, J., & Hahn, W. (2020). Are plant-based analogues replacing Cow's milk in the American diet? *Journal of Agricultural and Applied Economics, 1–18*. https://doi.org/10.1017/aae.2020.16

Stuart, T. (2009). *Waste: Uncovering the global food scandal*. WW Norton & Company

Taylor, C. A., Boulos, C., & Almond, D. (2020). Livestock plants and COVID-19 transmission. *Proceedings of the National Academy of Sciences, 202010115*. https://doi.org/10.1073/pnas.2010115117

Toensmeier, E. (2016). The carbon farming solution: A global toolkit of perennial crops and regenerative agriculture practices for climate change mitigation and food security.

U.S. Department of Agriculture. (2016a). *U. S. Agriculture and Forestry Greenhouse Gas Inventory, 1990–2013* (Technical Bulletin No. 1943). US Department of Agriculture. https://doi.org/10.15482/USDA.ADC/1264344

U.S. Department of Agriculture. (2016b). *U. S. Agriculture and Forestry Greenhouse Gas Inventory, 1990–2013*. U.S. Department of Agriculture, Office of the Chief Economist, Global Change Program Office. https://www.usda.gov/sites/default/files/documents/USDA_GHG_Inventory_1990-2013_9_19_16_reduced.pdf

U.S. Department of Agriculture Agricultural Marketing Service. (n.d.). *Research & Promotion Programs*. Agricultural Marketing Service. Retrieved May 5, 2021, from https://www.ams.usda.gov/rules-regulations/research-promotion

U.S. Department of Agriculture, & U.S. Department of Health and Human Services. (2020). *Dietary Guidelines for Americans 2020—2025*. U.S. Department of Agriculture. https://dietaryguidelines.gov/sites/default/files/2021-03/Dietary_Guidelines_for_Americans-2020-2025.pdf

U.S. Department of Agriculture Economic Research Service. (2020). *Meat supply and disappearance: Meat production, imports, total supply, exports, total and per capita disappearance with history*. https://www.ers.usda.gov/data-products/livestock-meat-domestic-data/livestock-meat-domestic-data/#All%20supply%20and%20disappearance

U.S. Department of Agriculture, National Agricultural Statistics Service. (2020). *Milk production (February 2020)*. https://www.nass.usda.gov/Publications/Todays_Reports/reports/mkpr0220.pdf

U.S. Department of Agriculture, National Agricultural Statistics Service. (2021). *Milk production (February 2021)* (ISSN: 1949-1557; Agriculture Counts). U.S. Department of Agriculture, National Agricultural Statistics Service. https://downloads.usda.library.cornell.edu/usda-esmis/files/h989r321c/br86bx43m/7p88d932r/mkpr0221.pdf

U.S. Department of Energy Office of Science. (2008). *Carbon cycling and biosequestration: Integrating biology and climate through systems science; report from the March 2008 workshop* (DOE/SC-108). U.S. Department of Energy Office of Science. https://genomicscience.energy.gov/carboncycle/

U.S. Environmental Protection Agency. (2020). *Inventory of U.S. Greenhouse Gas Emissions and Sinks 1990–2018*. https://www.epa.gov/ghgemissions/sources-greenhouse-gas-emissions

Verbrugghe, A. (2018). Different protocols and technologies employed to help pets lose weight. *Journal of Animal Science, 96*(Suppl. S3), 152. https://doi.org.ezproxy.library.wisc.edu/10.1093/jas/sky404.331

Waller, D. M., Meyer, A. G., Raff, Z., & Apfelbaum, S. I. (2021). Shifts in precipitation and agricultural intensity increase phosphorus concentrations and loads in an agricultural watershed. *Journal of Environmental Management, 284*, 112019. https://doi.org/10.1016/j.jenvman.2021.112019

Wattiaux, M. A., Uddin, M. E., Letelier, P., Jackson, R. D., & Larson, R. A. (2019). Invited review: Emission and mitigation of greenhouse gases from dairy farms: The cow, the manure, and the field. *Applied Animal Science, 35*(2), 238–254. https://doi.org/10.15232/aas.2018-01803

Weil, R. (2017). *Building soil organic matter: What, why, how?* 2017 National Conference on Cover Crops and Soil Health. https://www.sare.org/wp-content/uploads/Weil-BuildingSOM_WEIL_2017NatlConfCoverCropsSoilHealth.pdf

Winders, B., & Ransom, E. (2019). Introduction to the global meat industry: Expanding production, consumption, and trade. In *Global meat: Social and environmental consequences of the expanding meat industry*. MIT Press.

Wisconsin Farmers Union. (2020). *Supply management and how it can save our rural economy.* https://www.youtube.com/watch?v=hCD-prKQFeM

Wisconsin Milk Marketing Board. (2021). *Wisconsin Dairy Promotion.* Dairy Farmers of Wisconsin. https://www.wisconsindairy.org/Promote-Dairy

Chapter 6
The Miracle of Grass

W. Richard Teague and Steven I. Apfelbaum

Abstract Historically, nearly 40% of terrestrial ecosystems were grasslands and savannas that co-evolved with grazing animals. Historic wild herd numbers were probably only slightly lower than today's combined global herds of wild ungulates and domesticated livestock. Thus, modern degradation of critical ecosystem services in these ecosystems, including increases in atmospheric greenhouse gases caused by poor grazing and agricultural management, cannot simply be attributed to livestock numbers.

Livestock grazing management investigations indicate that emulating the fast and short grazing periods of co-evolved wild herds of native grazers, followed by long recovery periods, can provide significant benefits over continuous grazing in these ecosystems. Such benefits include improved water infiltration, decreased soil erosion, increased soil fertility, vegetative and ecosystem biodiversity, habitat for rare grassland birds, increased vegetation productivity, and increased soil organic carbon stocks, often to a meter depth in many areas.

Keywords Adaptive multipaddock grazing · Grassland soil carbon · Livestock and climate · Climate change · Grasslands · Livestock · Soil carbon

---

W. R. Teague
Rangeland, Wildlife, and Fisheries Management, Texas A&M AgriLife Research, College Station, TX, USA

Borlaug Institute, Vernon, TX, USA

S. I. Apfelbaum (✉)
Applied Ecological Institute, Inc., Juda, WI, USA
e-mail: steve@aeinstitute.org

D. Mayerfeld (ed.), *Our Carbon Hoofprint*, Food and Health,
https://doi.org/10.1007/978-3-031-09023-3_6

6.1 Introduction

6.1.1 What Is the Miracle of Grass?

The evolution of grasses and grassland ecosystems is an essential global example of how life has negotiated the thermodynamics and energetics of the earth. Grasslands were one of the ecosystems heavily involved in the sequestration and modulation of living carbon stocks, water cycles, and fostering life above and below ground — a miracle by any definition.

If the shrubs, trees, and early grass relatives had not transformed by shifting their growth centers vulnerable to wind, fire, and herbivory to below ground apices and stolons, this ecosystem would not exist as we know it. The evolutionary change to protect below ground growth centers on grass plants has profoundly reshaped the planet, resulting in the development of soils capable of supporting more productive human food systems, and soils that support micro-landscape and macro-regional and global water cycles. This evolutionary change to below-ground growth centers in grasses, and other adaptations resulted in highly resilient grazed ecosystems that support more animal biomass and sustain considerably higher levels of herbivory than different terrestrial habitats (Frank et al., 2002).

Grasslands have consequently become one of earth's largest, most widespread ecosystems, and the benefits provided are fundamental to humanity and life as we know it on earth. Many civilizations have failed to recognize this miracle. Where people have degraded the soils, this behavior stands as a perplexing misalignment between long-term human needs and contrived shortsighted economic systems, where the depletion of grasslands and soils is considered an economic externality. Civilizations exhibiting this behavior found themselves on shaky ground, and most no longer exist (Montgomery, 2012).

The miracle of grass is a shining example of an efficient thermodynamic system. The functions of long-lived, durable plants have responded to sunlight, water, and atmospheric and geological mineralization by capturing energy via photosynthesis. The coevolution of grasslands with soil fungi and microbes, plants, and various associated animal life forms resulted in the formation of carbon rich soils over the last 40 million years (Retallack, 2013). This vast mass of stored carbon, our grassland soils systems-- is a form of currency—and provides healthy controls on watershed function, land productivity, soil nutrient retention, food nutritional density, human health, and our economies. Land with poor, depleted soils and carbon stocks generally co-occurs with impoverished water supplies, human health, and human economies. Real wealth comprises the riches and potential value contained in the soil.

Grasslands, like all ecosystems, can self-propagate, self-regulate, and expand (Odum, 1997), as a living, growing example where nature exhibits the products of complex partnerships of life—where microbes, plants, wildlife, humans, soil fungi, and water cycles work together. This mutualism is the result of these groups of organisms adapting to tolerate and grow under heat and cold, drought and freezing, cycles of erosion, insect depredation and grazing, wildfires, and other environmental stresses.

Modern agriculture has unfortunately seriously threatened the grassland miracle. The ecological services provided by natural grasslands have been largely forgotten or ignored in the development of modern agriculture. As a result, diverse perennial grasslands have been degraded by poor grazing practices or replaced in many areas by annual monocultures that rely on tillage, as well as artificial props of chemical fertilizers and pesticides. These props have been made very accessible by crop subsidies, innovative equipment, and underpinning finance to provide loans for herbicides, seeds and fertilizers, and state-of-the-art equipment. At the hand of modern man, this technology, equipment, and financial support has duped consumers into thinking inexpensive food is what they need, without consideration of unintended consequences to ecological function and services. Technology, chemicals and biocides have replaced the naturally evolved adapted biota of many native grasslands, with the underlying presumption that this technological approach is better than the system nature developed over millions of years, where the judge of success was survival - not a bank loan provided to a farmer to stay afloat for another year, or decade.

Deep-rooted native grassland plants, microbes, and other life co-created the soils that supported immense herds of bison, caribou, wildebeest, antelope, and hundreds of millions of other grazing animals. They occupied the 3.5 billion hectares of rangelands, the 2–3 billion hectares of seasonally moist wetlands, and rich and fertile soils that are the modern-day crop production breadbaskets that feed humanity. These grasslands included semi-desert grasslands, coastal plains, fresh and saltwater marshes, the immense North American great plains and tallgrass prairie, Asian steppes, tropical and subtropical grasslands. Through the ecosystem service benefits they provide, evolved native grasslands have contributed to the nutrient-rich productive soils we rely on — a tangible reminder of why we must nurture the miracle of grass now and in the future.

6.2 The Stage Was Set 700 Million Years Ago

When the first plants ventured out of the ocean onto land, life on earth changed. Species of plants and life that relied on the molecular bond energy in sulfur and other compounds spewed from deep-sea vents, moved into coastal, brackish and freshwater and began competing for more readily available sunlight as a source of energy, which opened up tremendous opportunities and furthered the evolution toward terrestrial occupation. Primary carbon sequestration that occurred over millions of years included mineralization of carbonates, and deposition of inorganic carbon, which resulted in fossilized carbon in the form of limestone, often mixed with organic carbon matter, bodies of ancient sea animals and plants, deposited and also found now in shale (former mud), and slate. And, over eons, much of this morphed into fossilized forms that include the conversions into coal, oil, and undecomposed peat plant materials.

Geological evidence relating to the Eocene extinction about 34 million years ago suggests it was caused by a meteoric or comet collision with earth. This event coincided with the loss of diversity of most life forms on the planet. Subterranean species and groups appeared to have survived, including small surviving mammals protected below ground, or groups protected because of aquatic system occupation. Did this same event favor plant species with below-ground growing centers? Scientists exploring the nexus of archeological, paleo botanical, and geological evidence have indicated that something sparked the initiation of the grassland miracle at this time. This period (Retallack, 2013) coincided with the evolution of stoloniferous grasses, that invested in deeper rooting, more rapid carbon fixation and new metabolic pathways of water use efficiency through to evolution of the C_4 photosynthetic pathway used by warm-season grasses (Retallack, 2013). This photosynthetic strategy is more efficient in water and energy utilization than the older C_3 photosynthetic pathway. At this time, a counterpoint dance began when grazing wildlife developed the teeth to graze these grasses. In that dance, the grasses responded through time and developed phytoliths (silica stone cells) and other herbivory deterrents, including pathways that resulted in secondary metabolites to deter herbivory, such as oxalic acids. These compounds can impart unpleasant chemical tastes in above-ground plant tissue. The evolutionary dance continued as grazing herbivores evolved the ability to pass oxalic acids and other chemical herbivory deterrents in their urine. Oxalic acid becomes an inorganic soil carbon addition and needs to be understood in more detail but appears to be associated under very high grazing ungulate stocking rates with increases in soil inorganic carbon (Apfelbaum, unpublished data). This dance involved many other organisms, including perhaps hundreds of thousands of species and continuously evolving groups of organisms that we now call soil microbes, including bacteria, fungi, archaea, algae, soil insects, and many other organismal groups and plants.

The age of mammals exploded with the radiation of mammal diversity, especially those co-dependent on the grasslands. The evolutionary results of this process mainly endure to present day, with some stochastic events that recast the mammalian participants. Some Pleistocene giants, such as the enormous Pleistocene bison (*Bison latifrons*), beaver (*Castor spp.*), giant tree sloth (superorder *Xenarthra*) and others suffered extinction, while successor relatives have endured in the form of *Bison bison*, and modern beaver (*Castor canadensis*), all smaller in stature, and perhaps more numerous.

6.3 Humans Meet the Grassland Miracle

Grazing ecosystems of native grassland and savanna occupy approximately one-third of the earth's land area, and many are unsuited for crop cultivation because of precipitation, edaphic or topographic and geologic limitations (Delgado et al., 2011;

MEA, 2005). Rural and urban communities in these drier grassland and savanna areas, often with challenging soils and topography, rely almost wholly on grazing and browsing by domestic livestock and wildlife to support them and to provide essential ecosystem services to underpin their existence (CAST, 1999; Janzen, 2010; Ottoboni & Ottoboni, 2013; Ragab & Prudhomme, 2002). Many of these ecosystems are degraded primarily due to inappropriate land-use practices, negatively affecting the well-being of at least 1 billion people worldwide. Although soil and ecosystem degradation are acute in many of these regions (Jandl et al., 2014), evidence from land managers in many grazing ecosystems indicates that adaptive, goal-directed grazing and cropping management can cost-effectively reverse degradation (Savory & Butterfield, 2016).

Currently available genetic and archaeological evidence is supportive of a recent single origin of modern humans in east Africa (Liu et al., 2006) Depending on the accepted theory (see extensive literature by Richard Leaky, others), human evolution and radiation from Africa's Rift valley in Tanzania coincided spatially and temporally with the development of grassland ecosystems. It is very likely that the savanna systems were present where human evolution is thought to have occurred, may have been influenced by prehistoric humanity. Human fear and fascination with fire and need to find and harvest game to underpin their livelihoods and well-being led to using fire to clear lands and attract game to facilitate successful hunting. Deliberate fires perhaps also supported wind flow to reduce biting insects around habitations (Brain & Sillen, 1988). While lightning sometimes fired the grassland landscapes, human propensity for using this tool likely occurred at altered frequencies and intensities and directly shaped the patterns, diversity and composition, and perhaps even extent of grasslands.

For thousands of years humans relied on and managed grasslands to support wild animals they could hunt. Then, about 8–12 thousand years ago, in temperate and tropical regions humans started breeding and domesticating both plants and animals. By the beginning of the twentieth century only a few societies such as the Inuit in the Arctic relied primarily on hunting and gathering wild species. Around the world domestication of crops and livestock has led to increased human populations – and in turn to increased demand for plant and animal products.

In the past 150 years, agriculture has emphasized crop production and various strategies for increased yields, including genetic modifications associated with crop and livestock selection. This focus on planted crops has led to the large-scale replacement of the adapted, durable, and biodiverse native plant communities that once provided food and other ecosystem services. This included tillage and replacement with plantings of monocultures of plants, often annual or short-lived crop plants and fertilizer subsidy to support the higher yield expectations necessitated by the very cropping practices adopted. Agricultural development and colonization since 1800 also imposed a significant impact on land used for livestock production but in a way that runs counter to the historical relationship between land and grazers. In many parts of the world grasslands that were previously grazed periodically by nomadic herds managed by pastoral groups shifted to continuous grazing without recovery after grazing. This shift has caused over-consumption of the forage base by the livestock, inadequate recovery of above-ground plant material, and decreased root biomass and depth of forage plants. This overgrazing often results in increased bare ground, depletion of native plant biodiversity, a shift to more grazing tolerant, weedier species, and substantially reduced forage production and provision of essential ecosystem services.

The primary cause of ecosystem degradation in these ecosystems is continuous year-round grazing, which is commonly exacerbated by stocking at levels that exceed carrying capacity (Frank et al., 2002; MEA, 2005; Teague et al. 2013). However, having too few or no grazing animals can also be detrimental, as grasslands thrive and remain competitive under infrequent and light to moderate defoliation. They deteriorate in the absence of periodic disturbance in the form of fire, mowing, or grazing (Knapp, 1985). These periodic disturbances, followed by adequate recovery, are necessary to maintain ecosystem function in grassland ecosystems. Periodic disturbance facilitates the capture of solar energy via photosynthesis that drives the ecosystem functions; cycling of minerals through soil, plants and animals; and provides the biodiversity that enhances all ecosystem processes (Frank et al., 2002).

It is widely believed that stocking an area at levels higher than its current grazing capacity is the critical management factor that needs to be adjusted in order to sustain long-term profits while maintaining ecosystem function (Huffaker & Cooper, 1995; Kobayashi et al., 2007; Ritten et al., 2010). However, appropriate stocking rate alone does not prevent rangeland degradation (Jakoby et al., 2015; Müller et al., 2007, 2014; Savory & Butterfield, 2016; Teague et al. 2013); it is also necessary to time grazing and distribution of grazing effectively, because livestock in large paddocks exhibit repetitive use of preferred plants and patches. This behavior leads to overgrazing of preferred plants and patches and is a major cause of unwanted changes in rangeland ecosystems (Barnes et al., 2008; Massy, 2018; Norton, 1998, 2003; Teague et al. 2013).

Repeated consumption of preferred plants and patches often results in uneven impact, such that even at low stocking rates, localized degraded patches persist over time and progressively expand, eventually degrading the landscape (Bailey et al., 1998; Bullock et al., 1994; Fuls, 1992; O'Connor, 1992; Teague et al., 2004). Poor grazing practices lead to soil compaction and reduced infiltration rates, exacerbating the most limiting factor in most grazing ecosystems, which is plant-available soil water (Thurow, 1991). Runoff increases soil loss and facilitates nutrient movement by surface runoff, which can lead to the eutrophication and impairment of freshwater sources (Babiker et al., 2004; Burkart & Stoner, 2002; Gillingham & Thorrold, 2000; James et al., 2007; Sauer et al., 2000; Sharpley et al., 1994; Stout et al., 2000; Vadas et al., 2014; Webber et al., 2010). Soil erosion and nutrient losses are generally considerably less under rotational and adaptive multi-paddock (AMP) grazing than from continuously grazed pastures (Mathews et al., 1994; Park et al., 2017; Ritter, 1988; Sovell et al., 2000; Stout et al., 2000; Webber et al., 2010; Weltz et al., 2011). For long term sustainable use, managers must adopt grazing management practices that restore, or at least maintain, soil and ecosystem function and resilience (Havstad et al., 2007; Teague et al. 2013).

6.4 Ecology of Grazing Ecosystems

Paleo-geological data from the Oligocene and Miocene provide evidence that the evolution and spreading of first tufted then sod grasslands during the late Eocene and early Miocene displaced older vegetation types in semiarid to sub humid climatic periods. The global expansion of grasslands and associated soils co-evolved under changing environmental conditions as complex, dynamic ecosystems comprising co-adapted grasses and soil biota, the grazers, and their predators (Retallack, 2013). These ecosystems were dominated by large, mobile herds of large-bodied ungulates, and their grazing, along with fire and fluctuating climatic regimes, created a dynamic resilience of organisms that regularly respond to biophysical events. The resulting highly resilient grazed ecosystems supported more animal biomass and sustained higher levels of herbivory than other terrestrial habitats (Frank et al., 2002).

The key elements characterizing grazed ecosystems are spatial and temporal variation in plant diversity, forage supply, and dominance by large migratory herds of herbivores (Frank et al., 2002; McNaughton et al., 1989). These continually changing elements caused the grazing herds to move regularly to satisfy water and nutrient requirements, avoid sites fouled each day by urine and feces, maintain social interactions, and respond to influences of fire, predation, herding and hunting (Provenza et al. 2003a, b; Bailey and Provenza, 2008). Although grazing pressure under these circumstances was intense at some sites, because herds continually move, concentrated grazing in natural and traditional grazed ecosystems seldom lasts long, and grazed plants are typically afforded time for inter-defoliation recovery when herds move to new feeding grounds (Frank et al., 2002).

During these ongoing events, grazing herds impact many ecosystem processes and other smaller life forms that together often exceed the living biomass of the grazing ungulate herds. Nutrient concentrations are increased by stimulation of soil microbiota that contribute to soil function, as well as by the urination and defecation of the large animal herds (Frank & Groffman, 1998; Holland et al., 1992). These changes enhance mineral and water availability for soil microbes and plant roots, which in turn positively influences photosynthesis via improved plant nutrition to increase plant production compared to ungrazed areas (Bryant et al., 1991; Frank et al., 1998; Hamilton & Frank, 2001). Thus, periodic herbivory can increase plant productivity and soil organic matter, and can improve soil aggregation, soil moisture, and nutrients. In combination, these soil changes generate conditions that are more conducive for plant growth than for the development of chemical defenses by plants, and so enhance palatability and nutritive value of plants (Bryant et al., 1991; Coley et al., 1985; Provenza et al., 2008). Such effects of grazers on soil carbon and nutrient distribution have landscape-wide ecological impacts (Frank & Groffman, 1998) that are limited only by periods of low moisture or extreme temperature conditions that limit the function of the biological community.

Perhaps the most fundamental relationship occurred within the soil-plant root zone. There, plants and soil microbes have, over time, developed means of supporting one another that are mutually beneficial in many ways. These relationships include natural turnover of root, leaf, insect, animal and microbial wastes, and herbivore-deterrent chemical exudates, all of which become a desirable food source for microbes driving the essential process of nutrient cycling. A passive release of root exudates, especially exudates that contain sugary substances and amino acids, fosters a microbial concentration and activity massing around the roots that initiate complex, mutually beneficial exchanges between plants and associated mycorrhizal fungi and bacteria. The arbuscular mycorrhizal fungi (AMF) that are provided root exudate from associated plants provide access to water and minerals to their host plants by mineralizing elements in the rooting zone that would not otherwise be available to the host plants. Saprophytic fungi also play an essential role in boosting ecosystem function by breaking down and recycling dead plant and animal matter to enrich the rooting zone with chemical constituents used by plants (Bardgett & McAlister, 1999; De Vries et al., 2013). The symbiotic relationships legumes have with fungi and bacteria likewise benefit plants, microbes, and other life organized around the plant rooting zone.

Extensive grasslands are found on all continents except Antarctica, and on many near-continent islands. Modern-day grasses' methods of self-propagation occur through fragmentation of the stolons, rhizomes, seed production, and radial growth, and using grazing mammals for seed dispersal. These interactions among grazers, grasses, forbs, microbes, water systems, and soil nutrients helped build earth's rich grassland soil stocks (Retallack, 2013).

Grazers have teeth with specific incisor and molar adaptions for tearing tough plant tissues and grinding/chewing grassland species. Numerous other behaviors evolved, including herding behavior, which unites the senses of individuals into a super-organism's sensitivity to detect and avoid predators, and to take advantage of the gregariousness factor for increased breeding success and collective rearing of young.

6.5 Ecosystem Services

A tabulation of ecosystem services used to evaluate ranch sustainability and trends is presented in Table 6.1. Although the primary focus of this book is on climate-regulating gases, we cannot ignore the other ecosystem services needed to support a healthy planet and human population. It is critical to the understanding of the miracle of grass to explore the scale, nature, and magnitude of historical and present-day services of the grassland ecosystem. The human zeal for converting land and the ecosystems present to something else is truly amazing. So too is how we have done this with limited information and often as uninformed creators of change. Since the industrial period began, a massive process of resource extraction has supported rapid human population growth with commensurate consumption of resources. At the same time, humans have plowed up, deforested, drained, bulldozed, and covered over with pavement and homes very large portions of the former grasslands. The

Table 6.1 Parameters for evaluating increases in farm or ranch sustainability and trend

Attribute	Indicators	Comments
Economic		
Productivity	Sales of products produced	The capacity of the ranch business to provide the goods and services that sustain livelihoods and ecosystem function.
Flexibility	Ability to adapt to changes	Capacity to recover from climatic, market and social perturbations.
Stability	Secure land tenure, locally adapted livestock	Long-term maintenance of profitability and low annual variation in income.
Self-reliance	Amount of purchased inputs and subsidies, locally adapted livestock	Lower dependence on external inputs and subsidies indicates greater self-reliance.
Return on investment	Cash flow from an investment to the investor	Ratio of money gained or lost relative to that invested.
Social		
Equity	Total ranch AWU[a]/100 ha	System's ability to equitably distribute profits and cost intra- and inter-generationally. Gaspar et al. (2009) consider that the greater the amount of adequately rewarded labor employed by a ranch, the more equitable the business.
	Family AWU/100 ha	
	Permanent AWU/100 ha	
	Temporary AWU/100 ha	
Environmental		
Energy capture	Improved cover of green leaf, minimal bare ground	The amount of vegetation capable of capturing solar energy.
Water cycle	Plant canopy and litter cover of the soil	Ability of the area to capture incoming precipitation into soil and minimize erosion.
Vegetation	Dominance by perennial herbs, high seral herbaceous plants, less desirable plants	It is essential to manage for dominance by adapted native plants and communities that will promote the potential provision of goods and services possible for a given ecosystem.
Soil health and function	Soil carbon, nutrient status, soil formation or erosion	Management to increase soil carbon increases fertility and builds soil.
Mineral cycling	Rapid decay of fecal material and dead vegetation	Efficient cycling is essential to achieving productive potential and high biodiversity.
Biodiversity	Microbial, floral, faunal, and mixed livestock species	Essential to maintain adequate ecosystem function and the efficient provision of ecosystem services.
Resilience	Adaptability, transformability	Capacity of social-ecological systems to retain the same basic structure and function following disturbance.

López- Ridaura et al. (2002), Parsons and Pratt (1991), Savory and Butterfield (2016), Walker et al. (2002)
[a] Annual Work Units (Gaspar et al., 2009; López- Ridaura et al., 2002)

most productive historic grasslands have been converted to croplands. This conversion has resulted in a 70–90% depletion of historic soil carbon stocks in these cropped soils, along with decreases in water holding capacity, soil nutrients and subsequent crop nutrient quality and density. Annual row crops have also contributed to high levels of erosion, downstream flooding, and the deterioration of the watershed and overall landscape.

The depletion of vital elements such as the biodiversity of native plants and soil microbes has resulted directly in the reduction of former carbon stocks, with continuing declines in soil carbon as annually losses rather than gains accrue. This is land where shallow-rooted, often short-lived annual or biannual plant species replaced the long-lived, deep-rooted native plants that had microbial relationships developed over millennia. Carbon depletions may continue even after the most destructive of the land management practices have ceased, as biological diversity and cycles of heathy regeneration have been materially disrupted.

Development of grassland ecosystem services, and the legacy of delivery of such services over millennia has resulted in large grassland landscapes with some of the very highest soil carbon stocks on earth (second to only peatland carbon stocks which are found concentrated in a small percentage of the acreage, such as the pocosin wetlands in coastal North Carolina), as well as lush and productive soils. In addition to supporting a diversity of plants, animals, and plant-soil-microbial water relationships that modulate micro-climate, flooding, base flow, and so much more, grasslands regulate greenhouse gases through several pathways. Water vapor is the most prevalent greenhouse gas and, along with increases in carbon dioxide, has the most significant controlling influence on the climate of earth (Ferguson & Veizer, 2007; Pokorný et al., 2010; Veizer et al., 2000). Methane (CH_4) is more than 21 times more potent as a greenhouse gas than CO_2 but short-lived in the atmosphere. Oxides of nitrogen are also potent climate forcers. Grasslands affect all these critical greenhouse gases.

Decomposition of soil organic carbon, much like the decomposition of plant matter in a compost pile, mainly results in the liberation of carbon dioxide and water vapor, both being waste products of the soil bacteria, fungi, and insects involved in the process. Facilitating the entry of oxygen to these decomposers hastens the decomposition process. Therefore, plowing a native grassland can result in a quick 40% decline in soil organic carbon, and over time, regardless of the plants grown on the land, some significantly lower level of soil carbon (organic matter) can be observed under this continuous soil disruption process. This reduction in soil carbon has been documented to be associated with reduced rooting depth of stressed plants subjected to continuous grazing, and reduced microbial mass. Also, the addition of inorganic nitrogen fertilizer further hastens this organic matter oxidation, as increased nitrogen supports more bacteria that consume and break down the carbon from dead plant, microbe and animal matter. This increased decomposition results in more carbon dioxide and water vapor emissions.

Nitrous oxide (N_2O) can be produced during nitrification under aerobic conditions or during denitrification under anaerobic conditions and increases as soil moisture increases due to reduced diffusivity, the rate at which gases or liquids can move through the soil. This decreases O_2 availability. As temperature increases, so does respiration, also resulting in decreasing soil O_2 levels causing increasingly anaerobic conditions, which in turn increase N_2O emissions (Smith et al., 2003). Consequently, the amount of N_2O emitted depends substantially on soil structure and the wetness of the soil.

In addition, aerobic soils are an important sink for CH_4, and upland native grassland sites are generally aerobic and operate as weak sinks of CH_4 emissions (Saggar et al., 2007). They are net sources of CH_4 when anaerobic, such as when wet enough to reduce gas diffusivity. Soil temperature has a small impact on methane flux unless soil moisture is low enough to reduce microbial activity that is necessary to oxidize CH_4 (Smith et al., 2003).

Planned adaptive multi-paddock (AMP) grazing management was specifically designed to emulate evolved grazing ecosystem processes (Savory & Butterfield, 2016) and regenerates ecosystem functions degraded by the negative impacts of continuous gazing (Teague et al. 2013). AMP grazing management of herbaceous plants aims to maintain sufficient litter and plant cover of the soil to provide optimal microbial function, maximize rainfall infiltration, and ensure rapid rates of nutrient cycling and nutrient retention to enhance productivity and soil organic matter. These protocols manage for high biodiversity and optimum plant species composition to capture maximum sun energy via photosynthesis to facilitate optimal levels of key ecological functions and have demonstrated the potential to substantially improve ecosystems service outcomes relative to the most commonly used grazing management of moderate (MC) and heavy continuous (HC) grazing (Teague et al. 2011). AMP grazing results in lower N_2O emissions and is a larger CH_4 sink than continuous grazing (Dowhower et al. 2019) and stores higher levels of soil organic carbon than continuous grazing. In fact, AMP grasslands can store more carbon equivalent units in the soil than is emitted by respiration of soil biota or via ruminant emissions (Liebig et al., 2010; Mosier, 2020; Rowntree et al., 2016; Stanley et al., 2018; Wang et al., 2015). This is a major contributing factor to AMP providing a greater GHG sink because CO_2 emissions are lower from soils with higher carbon content and N_2O emissions are also decreased at higher levels of soil carbon (Gelfand et al., 2015; Ruser et al., 2006).

Most management of plants and soil on agricultural land is the opposite of how the historic grassland miracle evolved and can be maintained. The degraded ecosystems we have now cannot deliver the essential ecosystem services from both the farmland and grazing land that are essential to human well-being, as a consequence of how they have been managed. On farms, plowing, fertilizing, and biocides to kill perceived pests, as well as use of single species annual crops with shallow roots, leave only a small percentage of the field covered with living green plant tissue or crop residues, and these crop fields are now excessive emitters of greenhouse gases. On grazing lands, conversion to pastures with shallowed rooted, low diversity non-native plant species and poor grazing management create an ecologically inefficient grazing ecosystem.

6.6 Wild Ruminants Versus Livestock

What we are learning is that changes in total animal populations may be less significant than the nature of the shifts in livestock grazing practices. During the past century the world has experienced declining wildlife grazing on native grasslands and in savannas, to a condition where domesticated livestock now far outnumber wild animal stocks in most regions. Perhaps most importantly, these new burgeoning populations of livestock are primarily grazed in cultivated, tame pastures of fertilized, planted grasses and legumes. Many of these pastures are grazed without any recovery time and subjected to inputs that degrade soil biological functions, such as inorganic fertilizers and pesticides, and low plant diversity. The use of single species or low diversity plant mixes with high proportions of leguminous plantings that fix nitrogen, contributes directly to elevated nitrous oxide (N_2O) emissions, and drives soil carbon ratios down because of the decomposition of organic carbon in highly nitrogen skewed environments.

Recent decades have seen an increase in industrial feedlot livestock fattening operations (also known as concentrated animal feeding operations or CAFOs), resulting in the disposal of massive manure waste streams from these industrial farms. Erosion of manure nitrogen and phosphorous, and organic solids into the nation's waterways creates anaerobic conditions documented to emit extreme levels of N_20 and CH_4 (de Vries et al., 2013; Morriën et al., 2017) and cause ecological dysfunction of these waterways downstream with severe impacts on environmental and human health.

Hristov (2012), evaluated populations of bison and other ruminants prior to the spread of Europeans in North America (pre-1830's) and estimated that of all ruminants on the planet, historic U.S. bison herds represented the largest ruminant source of CH_4. Now the largest wildlife emitter in U.S. is white-tailed deer, and even their populations are lower than they were before European settlement. Hristov estimated that during pre-settlement times, wildlife in the area that is now the United States emitted roughly 86% as much methane per year as all farmed ruminants in the United States emitted in 2010. There is considerable uncertainty about the size of the bison herd before European contact, and under the highest bison population estimate of 50 million animals, historic wildlife may have emitted up to 23% more CH_4 than current domesticated ruminants, while under Hristov's lowest population estimate pre-European wildlife only emitted roughly 55% as much methane as livestock in 2010. Current populations of bison and other wild ruminants are only a fraction of what they were before 1800 and so emit less than 5% as much methane as domesticated livestock.[1]

[1] In 2011, USEPA estimated agricultural activities emitted 6.3% of total national emissions of 419 vs. 6633 Tg of CO_{2eq} yr.$^{-1}$. Methane from ruminant enteric emissions was estimated at 139 Tg of CO_{2eq} yr.$^{-1}$, about 20% of the national CH_4 emissions, but only 2.1% of the total national GHG emissions for 2009. In 2011, estimated emissions from bison, elk, and other ruminants were 0.28 Tg year^{-1}

These historic ruminant population estimates help set some broad boundaries around what total ruminant populations might be sustainable for North America, both in terms of net greenhouse gas emissions, and to restore the range of ecosystem services that grasslands grazed by bison, elk, and other wild ruminants provided before Europeans fenced and plowed the landscape. Even Hristov's low historic bison population estimate suggests that the U.S. could manage about half the current population of ruminant livestock in a way that significantly increases soil carbon, maintains current wild ruminant populations, and improves water quality and biodiversity. Under Hristov's middle estimate the U.S. could retain 75% of its livestock, double the current population of wild ruminants, and still meet climate and other environmental goals, with proper management.

The Hristov (2012) study was designed to be a conservative estimate at the time. We believe, with present-day updated understanding, that the actual methane emissions from wild ruminants before 1800 may have been lower than Hristov's model suggests for the following reasons:

1. Unlike domestic livestock receiving supplemental feed, wildlife experience a 3 month winter season through most of the bison's range in the U.S., when forage consumption and quality are both greatly decreased. Because enteric emissions are correlated with the quantity of feed consumed, the historic estimated CH_4 emissions from bison and other grazers in northern temperate and montane regions, would likely have been significantly reduced below an annual average emission rate for at least 3 months annually in winter. During summer hot, dry periods forage availability may likewise be reduced, so for more than a quarter of the year CH_4 emissions for wildlife could be expected to be well below the average emission levels that Hristov derived from measurements of emissions from penned livestock.
2. For purposes of standardization, Hristov used livestock CH_4 emissions from experimental lab settings where livestock are both far more docile, and are more accustomed to being penned up and eating the pelletized alfalfa used in their study. While standardization for purposes of testing seems reasonable, because of the paucity of the wildlife emissions data in general and near zero data from wildlife in their native habitat, there is considerable uncertainty about the general conclusions. Wild ruminants are far more active and consume much more diverse feed than that used to generate Hristov's CH_4 emission estimates.

   In recent native grassland studies early research has suggested that there may be significantly reduced CH_4 emissions (Apfelbaum, conversations with Dairy researchers) under a higher and balanced omega 3–6 fatty acid profile for dairy cows, such as one might find on healthy native grasslands. These findings suggest that wildlife, including bison, are unlikely to have year-round emissions similar to cattle, as their emissions on native range would result in the consumption of this more balanced fatty acid profile. Any study comparing essentially a feedlot emission level for cattle is not representative of the potentially much lower emissions that we believe would be associated with wildlife on native range. Is the reduced emission 30% or 80% or 100%? We don't know. In both cold and warm regions of the world, many plant species in native plant communities have adaptions that appear to be strongly associated with the presence of elevated levels of omega 3 fatty acid chemicals. Most domesticated pasture plants do not have these adaptations, and as a result, most domesticated livestock do not have the correct balance of these fatty acids to promote reduced CH_4 emissions. On the other hand, wild ruminants might also have higher enteric methane emissions sometimes because of lower forage quality at certain times of the year.

Further insights on these preliminary observations awaits the research findings from the dairy industry and others engaged in this research. Evidence of farmer interest in the potential for this effect to be used as a management strategy to reduce GHG emissions may now be found in the supplements being fed to dairy cows and other livestock in an attempt to reduce enteric emissions, including flax seed oil and other fatty acid oil supplements with a balanced Omega 3 and 6 fatty acid profile.

3. In healthy soil systems and under native and non-native grasslands using AMP grazing, it appears during the growing season in real time that soil methanotrophs may sequester emitted livestock CH_4 (Dowhower et al., 2019). While we are awaiting final sequestration rates and volumes, from a carbon 13 isotope pulsed study, this effect suggests that all estimates in this paper again may be significantly inflated for wildlife during the growing season periods. Currently, we do not know if the reduced emissions are 30% or 80% or 100%.
4. Globally, African, Asian, and North American ruminant wildlife numbers are significantly reduced from all documented historical populations. Saiga antelope (*Saiga tatarica*) numbers on the Asian steppes, have declined by millions of animals as have many other wildlife - caribou, bison, wildebeest, elephants, rhinoceri, most antelope, from historical levels. At the same time, because of drought incidence, domesticated livestock national herd sizes have been at their all-time lows in recent years. These documented declines are well beyond the scope of this paper and are not further described here. Total cattle numbers in the U.S. went from a peak of more than 130 million in the 1970s to fewer than 100 million since 2010 (U.S. Department of Agriculture Economic Research Service, 2021). Some wild ruminant populations are down by 95%, and some nearly extinct species, such as saiga antelope, have decreased even more significantly. These trends may suggest that when the decline of wild ruminants is taken into account, as well as decreases in U.S. cattle numbers, the causes of CH_4 level increases in the atmosphere need to be examined more comprehensively.
5. Taken together, these points suggest that existing concerns from IPCC and others about global ruminants and CH_4 are important to monitor, but it may be more helpful to focus on improving grazing management than to attempt to greatly decrease ruminant numbers, especially with the potential of improved AMP grazing and low disturbance cropping to restore soil health and move CO_2 from the atmosphere to the soil (Teague et al., 2016; Dowhower et al., 2019).

6.7 Managing Grasslands to Restore the Grassland Miracle

In many countries, leading conservation farmers have used adaptive multi-paddock (AMP) grazing to achieve superior soil health, vegetation, livestock production, and financial results (Teague et al. 2013; Savory & Butterfield, 2016; Teague & Barnes 2017). Similar positive resource and economic outcomes have been obtained by scientists who have studied the subject when research:

(i) was conducted at the scale of ranching operations;
(ii) was managed proactively as growing conditions changed to achieve desired ecosystem and production goals;
(iii) measured parameters indicating change in ecosystem function and not just production parameters; and
(iv) allowed sufficient time to elapse for management of the alternate grazing treatments to produce measurable differences resulting from the additive effects of positive impacts and many changes in factors like weather that influence responses over many years.

While many grazing studies in the scientific literature have concluded that rotational grazing is no better than continuous light grazing (Briske et al. 2008, 2011), these results came from rotational grazing management studies that were poorly designed and managed. These studies differed significantly from management used by successful proponents of AMP grazing, and thus they did not reflect the successes achieved with AMP on commercial ranches (Savory & Butterfield, 2016; Teague et al. 2013). The design and management of experiments profoundly impacts the results obtained, with favorable outcomes achieved by managing for specific goals, including soil health (Teague et al. 2013). Grazing practices extend on a continuum from overgrazing that results in soil carbon loss and even erosion to intensively managed grazing systems that improve both range health and financial results when management is deliberately designed to optimize environmental and economic outcomes (Teague & Barnes 2017).

The biggest limiting factor in most grazing land ecosystems is not the amount of rainfall but the amount of rainfall infiltrating the soil and how long it remains there (Thurow, 1991). But this is not the only important ecosystem function. Ensuring optimal ecosystem function requires efficient solar energy capture; effective water infiltration and retention; soil organic matter accumulation and retention; efficient nutrient cycling; and ecosystem biodiversity to facilitate these functions (Teague et al. 2013). Soil health is fundamental for ecosystem function, and 90% of soil function is mediated by microbes, with a mutual dependency among microbes, plants and animals. Plants enable microbial life and benefit in return from nutrient release through the synergistic interdependence among archaea, bacteria, fungi, and other microbial and eukaryotic species. How we manage plants in grazing or cropping ecosystems is critical to maintaining or regenerating full ecosystem function. The major portion of energy required to facilitate ecosystem functions comes from plant photosynthesis converting solar energy into carbohydrates that provide the energy for total community function.

The functions provided by networks of soil organisms include improving soil aggregation, aeration and water holding capacity (Altieri 1999); stabilizing soil (Van der Heijden et al., 2008); improving nutrient acquisition and retention (Green et al., 2008; Khidir et al., 2008); cycling nutrients to improve nutrient availability (Barrow et al., 2008; Rodriguez & Fraga, 1999); enhancing tolerance to biotic and abiotic stress (De Vries et al., 2012); and buffering the impact of environmental factors on plants (Van der Heijden et al., 2008). Arbuscular mycorrhizal fungi (AMF) are keystone species in terrestrial ecosystems, particularly grasslands, as they maintain plant diversity, mediate interactions among plants and other microbes, and positively impact plant photosynthesis (Averill et al., 2014). Plants increase photosynthesis in symbiosis with AMF and legumes for a dual association with rhizobia and AMF that enhances photosynthesis by 50% on average (Kaschuk et al., 2009). Arbuscular mycorrhizal fungi contribute directly to the soil organic matter pool, and through secretion of soil glycoproteins increase water stable soil aggregates that enhance soil water infiltration and aeration vital to ecosystem function (Rillig 2004).

Management decisions have important consequences for how grasslands support profitable operations, sequester carbon, and provide other ecosystem services. There are notable examples of management approaches that result in the restoration of degraded grasslands, especially where ranches are managed to achieve resource conservation goals (Teague et al., 2016; Teague & Barnes 2017). Improved management, such as AMP, has been shown to decrease bare ground, restore productive plant communities, increase water infiltration rates and soil water storage capacity, increase fungal to bacterial ratios and increase soil carbon, thus restoring rather than degrading overall ecosystem function (Delgado et al., 2011; Teague et al. 2011, 2013).

Positive shifts in soil microbial and biological community composition, carbon cycling and nitrogen cycling following improved management are strongly related to the restoration of soil carbon and fertility in grasslands (Altieri 1999; Van der Heijden et al., 2008; Nielsen et al., 2011; De Vries et al., 2012; Dowhower et al. 2019). At high microbial densities and species biodiversity, the expression of specific genes contributes to enhanced symbiotic interactions by microbiota to produce enhanced outcomes. This phenomenon is known as quorum sensing (Nealson et al. 1970). As management to restore soil function progresses, below ground microbial networks expand and fungal composition shifts, increasing the efficiency of nutrient cycling and carbon uptake in the soil (Morriën et al., 2017; Ngumbi & Kloepper, 2016; Slade et al., 2016). It is also possible to adjust grazing and associated farm management to optimize the benefits provided by other key organisms that have a strong influence on ecological function, such as dung beetles and earthworms (Herrick & Lal 1995; Richardson & Richardson 2000; Wardle & Bardgett 2004; Blouin et al., 2013).

The best benefits from improving grazing management have been achieved using practices specifically designed to enhance soil health and ecosystem function.

Fortunately, these practices also increase profitability, and these leading farmers have made substantial improvements in ecosystem function; plant species composition and productivity; soil carbon and fertility; water infiltration, and water-holding capacity; biodiversity; wildlife habitats; and profitability (Teague et al. 2013; Teague & Barnes 2017). Their method is to rotate a single herd among many paddocks, with short grazing periods and long recovery periods, adaptively moving the herd as indicated by residual biomass or other management elements such as growing conditions. It is becoming increasingly clear that the key to sustainable use and recovery from degradation involves using well-planned and adaptively managed multi-paddock grazing management protocols. These protocols include adjusting stock numbers to match forage biomass to achieve desired resource and financial goals (Earl & Jones, 1996; Jacobo et al. 2006; Provenza 2008, Ferguson et al. 2013; Teague et al. 2013; Jakoby et al. 2014; Martin et al., 2014; Müller et al., 2014; Jakoby et al., 2015; Teague et al., 2015; Savory & Butterfield, 2016; Wang et al. 2016).

Research that has followed protocols used by farmers to provide a high level of desired resource and economic improvements has substantiated farm-based results. Using a spatial simulation model, Jakoby et al. (2014) determined that appropriate grazing with a large number of paddocks per herd that included short grazing periods and adequate recovery facilitated resource improvement and gave the best financial results. However, economic risk was decreased only when management adjustments accounted for paddock forage quality and seasonality over the modeled landscape (Jakoby et al., 2015). Similarly, the model of Teague et al. (2015) found that too long a period of grazing or too short a recovery period resulted in more mediocre animal performance or plant recovery, with negative economic consequences, as acknowledged by experienced consultants working with farmers (Walt Davis; Dave Pratt pers. comm.).

Modeling of adaptive stocking with many paddocks per herd was found by Jakoby et al. (2015) to be less sensitive to overstocking than continuous stocking of one large pasture or long rotations among a few paddocks. Conversely, advantages of AMP over continuous grazing are less critical at low stocking rates but become increasingly crucial as stock numbers increase, improving net economic returns. Modelling by Wang et al. (2016) concluded that, at the scale of commercial ranches, AMP grazing with short periods of grazing and sufficient periods of post grazing recovery improved grass composition and productivity, as well as livestock dry matter consumption relative to continuous grazing, especially with higher stocking rates and lower initial standing crop and forage composition. However, the advantages of AMP grazing are less evident with favorable rainfall conditions, light stocking, low levels of undesirable plants, and short recovery periods. With their spatial model, Jakoby et al. (2015) identified several low-risk management choices that are viable with continuous grazing or few paddocks per herd, but they require relatively low stocking rates that result in low productivity and economic returns. In contrast, multi-paddock grazing using large paddock numbers, short graze periods, and long rest periods maintains or improves resource condition under both low and high stocking densities, resulting in superior economic outcomes, lowered income variability, and a higher likelihood of attaining a minimum income goal (Teague et al. 2011; Teague et al., 2015; Barnes et al., 2008).

6.8 Improving Grazing to Decrease the Carbon Footprint of Agriculture

One of the significant concerns in grazing land ecosystems is the substantial quantity of greenhouse gases emitted by ruminant livestock (Capper, 2012; Capper & Bauman, 2013; Eshel et al., 2014; Ripple et al., 2014). Although many scientists have concluded that ruminant production systems are an unusually large source of GHG emissions, others have found it is possible to convert many ruminant-based production chains into net carbon sinks by changing management (Rowntree et al., 2016; Teague et al., 2016; Wang et al., 2014; Wang et al., 2015). Previous assessments of capacity for CH_4 uptake in grazed rangeland ecosystems have not considered improved livestock management practices and thus underestimated potential for CH_4 uptake (Delgado et al., 2011; Rowntree et al., 2016; Stanley et al., 2018; Wang et al., 2014).

As soils can be a significant sink of carbon depending on management practices (Conant et al. 2001; Liebig et al., 2010; Teague et al. 2011; Machmuller et al., 2015), soil carbon dynamics are an essential component of calculating accurate beef life-cycle-assessments (LCAs) (Rowntree et al., 2016; Stanley et al., 2018; Teague et al., 2016; Wang et al., 2015). When conducting LCA assessments on emissions from ruminants in a food production chain, it is essential to include all elements in the chain influencing the net carbon footprint in the whole system under review (Teague et al., 2016). This includes accounting for beneficial ecosystem services such as those from carbon sequestered in grazing ecosystems that well-managed grazing systems can provide (e.g., Teague et al. 2011; Machmuller et al., 2015).

Under appropriate management, a grass-fed ruminant model can not only provide for the food requirements of the livestock but also be ecologically beneficial and regenerative. Ruminants grazing perennial grasslands that do not receive any negative management or degrading inputs that may reduce soil carbon levels and sequestration, have a smaller carbon footprint (Teague et al., 2016). Liebig et al. (2010) found that cattle on continuously grazed native rangeland in the Northern Great Plains of U.S. had a net margin of carbon sequestered over greenhouse gas emissions from cattle and soil of 0.607 Mg CO_{2eq} ha^{-1} $year^{-1}$ for moderate stocking, and 0.134 Mg CO_{2eq} ha^{-1} $year^{-1}$ for heavy stocking (using Natural Resource Conservation Service protocols).

Where the grazing management has resulted in improved ecosystem function and productivity, as reported above, the margin of increased carbon sequestered over ruminant emissions can be considerably higher. In southern tallgrass prairie in Texas, Wang et al. (2015) report a net increased carbon sink of 2.0 Mg carbon ha^{-1} $year^{-1}$ when converting from heavily stocked continuous grazing to AMP grazing at the same stocking rate, and net sink of 1.7 Mg C ha^{-1} $year^{-1}$ following conversion from heavy to light stocking with continuous, season-long grazing. Similarly, Rowntree et al. (2016), working with beef cattle on cultivated perennial pastures in Michigan, report soil carbon gains of 3 Mg carbon ha^{-1} $year^{-1}$ with AMP grazing under rainfed or irrigated conditions. For a net-zero GHG footprint, sensitivity analyses indicated that soil in the rainfed and irrigated AMP grazing systems would need to sequester 1 and 2 Mg carbon ha^{-1} $year^{-1}$, respectively.

Most cattle in North America are fattened in feedlots on grain-based feeds, and proponents of this finishing method claim that it results in lower GHG emissions per kilogram beef produced and a smaller carbon footprint. This conclusion is based on the fact that grain finishing reduces the overall production time to slaughter and therefore the lifetime enteric fermentation, relative to grass-based finishing (Stackhouse-Lawson et al. 2012; Capper, 2012; Capper & Bauman, 2013; Eshel et al., 2014; Ripple et al., 2014). However, these authors do not consider the full food-chain carbon footprint of grain-based finishing. To do so, they would need to account for the differences in GHG emissions associated with the production of grain-based feeds compared to pasture, including inorganic fertilizer production and nitrous oxide emissions, soil erosion, and other elements adding to carbon footprint levels (Teague et al., 2016). Ruminant dams and their offspring spend most of their lives on perennial grass. During this time the carbon sequestered by the grassland they graze can exceed their emissions under appropriate management (Rowntree et al., 2016; Stanley et al., 2018; Wang et al., 2015). This must be included when calculating the complete carbon footprint through any food-chain option being assessed.

In developed countries that routinely finish ruminants on grains, another factor that can decrease the carbon footprint of any production chain is the use of regenerative cropping practices with a negative GHG footprint (net carbon sink) to produce the grain (Aguilera et al., 2013; Gattinger et al., 2012). Such modification of agroecosystem production systems and conversion to regenerative cropping and AMP-based grass-finished livestock would also provide other significant ecological benefits (DeRamus et al., 2003) as outlined earlier in this document. Also, in the U.S. and Europe supplies of human food products could be increased by 70% if crop production currently used for animal feed and other uses such as biofuels were instead used for human food products, providing sufficient resources for billions of people (West et al., 2014).

Even simple grazing, when you look after the grass reasonably well, will put more carbon in the ground than the emissions emitted by the cattle grazing—up to about three or four times as much. In the more sophisticated grazing systems we have been studying, there is an order of eight times as much carbon dioxide equivalents being sequestered into the soil as is being emitted by the cattle. This only pertains if the livestock are on perennial grass. As soon as you take the animals off permanent pasture and feed them corn, their net emissions increase substantially. Industrial cropping methods have a large carbon footprint due to the amount of fossil fuels used in cropping and the very high soil erosion that occurs with such means of production (Rowntree et al., 2016; Stanley et al., 2018; Teague et al., 2016; Wang et al., 2015). When you feed cattle corn, they inherit the corn's substantial carbon footprint. And that's one of the reasons why grass-fed and grass-finished beef has caught people's attention. Being fed only on perennial grass without fossil energy-based inputs creates a robust carbon negative production system.

On perennial pastures the ground is covered year-round, reducing soil erosion and enhancing soil microbial diversity, biomass, and functionality. Carbon emissions from high input cropping agriculture are potentially high, and permanent pastures with low inputs are the best means of reducing it. We still need crop production, but with livestock integrated in the cropping system you can grow crops with no-tillage and cover crops that result in a negative carbon footprint, as folk like Gabe Brown and the Rodale Institute have shown (Aguilera et al., 2013; Delgado et al., 2011; Gattinger et al., 2012). With a robust negative carbon footprint from cropping using organic regenerative methods, as well as from a grazing component, then you can create a carbon-negative agriculture.

However, care has to be taken not to use any management that can degrade ecosystem function. For example, although many people assume that organic farming methods are good for the ecosystem, it is crucial to make sure that all the practices used in each case enhance soil and ecosystem function. All management used should be regenerative and organic. Many people who claim to be farming organically use bad management, such as excessive tillage, short rest periods, or biocides that kill beneficial organisms like dung beetles. Using only regenerative practices is essential as the use of degrading tillage, chemicals, and biocides will increase the carbon footprint. Farming using regenerative organic guidelines can result in a strong carbon negative footprint.

6.9 Distractions and Myths

Another amazing ecosystem service grasslands provide is grass-fed meat. It is a high-quality food that supplies essential macro- and micro-nutrients when produced in the way native grasslands were historically grazed by wildlife and livestock before modern agriculture disrupted grassland ecological processes. Food products from grazing animals have higher quality protein than those from plants. Ruminants often provide a cheap source of healthy protein and fat for humans; increase bioavailability of essential minerals; degrade anti-quality plant components such as phytates; and provide vitamin B12, which is not available from plant sources, as well as other vitamins. Protein food from grass has the best omega 3–6 ratio for human health (CAST, 1999; Ottoboni & Ottoboni, 2013; Mottet et al., 2017). Details matter; while grain and grass fed beef contain many of the same nutrients, mounting evidence suggests the proportions and ratios of key nutrients differ. Even the tertiary structure of the nutrient molecules themselves may vary in significant ways (e.g. not everything measured and called a protein, or an Omega 3 or 6 fatty acid is molecularly identical, and nutritional tests are often not designed to decipher molecular differences within a category or family of molecules). In an age where opportunities are always being thought up to increase profits, new food products and products used in food production chains arise without due consideration of unintended consequences. Investment regularly occurs using slick marketing and advertising to create markets to yield quick profits. Invariably not enough attention is paid to possible negative impacts on human or ecological health. Therefore, it is imperative for the sake of recovering the grassland miracle and all the ecosystem service benefits it provides, to not get distracted by the latest and greatest technological food system gimmicks that may cause more harm than good.

Food supply chains are rapidly being disrupted by ideas that could help restore the grassland miracle and others that surely don't understand negative impacts on human health and the global environment. In descending order we believe the following innovations likely have damaging implications for human and environmental health:

1. *Faux Meats* - Meats derived from industrial agriculturally produced grains (corn, peas, beans, etc.) carry the baggage of the significant CO_2 and N_2O emissions associated with the agricultural crop: inorganic fertilizers, biocides such as glyphosate, hauling, processing and fabrication life cycle costs. Faux meats add additional GHG emissions to the carbon footprint to equivalent meat products because of the post-processing emissions of the new faux products above and beyond the emissions associated with on farm agricultural production. The faux meat products also have great exposure to harmful biocides like Glyphosate as the crops used to produce them have more of these biocides in their production chains (Rodgers & Wolf, 2020).
2. *Enhanced Feedlot and Conventional Finishing of Meats* - Details matter here. Ineffective manure processing leads to higher ammonia, ammonium, CH_4, CO_2, and nitrogen oxide emissions under feedlot feeding, especially when the usually externalized emissions from the mining of oil and inorganic fertilizer progenitors, incipient soil erosion under row crop farming practices, all-grain handling, hauling, processing, and post-processing for feed and then meat products are accounted for.
3. *Grass-fed and Grass Finishing Operations* - Details matter here also. In native range and using AMP grazing (high density, short rotation, and appropriate long recovery periods) compared to high stock rate conventional continuous grazing or even few paddock rotational grazing, with short recovery periods, produce significantly different greenhouse gas emission profiles. AMP grazing, we believe, may have the lowest GHG emission profiles, primarily where grazing occurs in native plant dominated rangelands rather than tame, planted forage pastures. We also have evidence that grass finishing on continuously grazed paddocks has significantly higher GHG emissions than AMP grazed and finished (Dowhower et al. 2019). However, both continuous and AMP grazing have lower GHG emission profiles than the finishing and product protocols listed above (e.g. faux meat)(Wang et al., 2015; Rowntree et al., 2016; Stanley et al., 2018; Dowhower et al. 2019).

6.10 Conclusions

Details matter! There may be few ways to use a hammer as a tool, but each carpenter has a different style of swinging the instrument, and some are more adept at reliably hitting the nail on the head. The human invention of managing land, domestication of livestock and "harnessing the miracle of grasslands" to contribute to meeting our nutritional and caloric energy needs, is a relatively new invention with many moving parts and myriad ways to exercise our human ingenuity. The outcomes vary - the health and quality of the land, the livestock, and our dependence on the details of the management decisions we make.

A human life span is short, certainly in contrast to the life of ecosystems and the complex relationships found in an ecosystem. We may think we know the best management practices, but without any reference point to use, besides the dollars or beef pounds generated per acre, we create behavior strongly driven by perhaps the incorrect interpretations of what is working or not. We have unfortunately learned that behaviors become comfortably entrenched around legacy thinking and expectations, or through perceptions of peer pressure, or financial constraints. Because details matter - objectivity is essential.

We argue in this paper that we cannot truly understand the existing conditions and challenges of climate change for animal agriculture without understanding the historic evolutionary drivers that shaped the grassland ecosystems. Without this context, we can neither successfully create a healthy future, nor help others do so on their land and in their lives. Appreciating the miracle of grass, and the soils in which it grows, and the functional attributes of grassland ecosystems—microbes, plants, insects, grazers, and how they developed, is essential to understanding the responsibility we now have, to restore and steward the miracle of grass into the future.

References

Aguilera, E., Lassaletta, L., Gattinger, A., & Gimeno, B. S. (2013). Managing soil carbon for climate change mitigation and adaptation in Mediterranean cropping systems: A meta-analysis. *Agriculture Ecosystems and Environment, 168*, 25–36.

Altieri, M. A. (1999). The ecological role of biodiversity in agroecosystems. *Agriculture, Ecosystems & Environment, 74*, 19–31. https://doi.org/10.1016/B978-0-444-50019-9.50005-4

Averill, C., Turner, B. L., & Finzi, A. C. (2014). Mycorrhiza-mediated competition between plants and decomposers drives soil carbon storage. *Nature, 505*, 543–545.

Babiker, I. S., Mohamed, M. A. A., Terao, H., Kato, K., et al. (2004). Assessment of groundwater contamination by nitrate leaching from intensive vegetable cultivation using geographical information system. *Environment International, 29*, 1009–1017.

Bailey, D. W., Dumont, B., & M.F. (1998). Utilization of heterogeneous grasslands by domestic herbivores: Theory to management. *Annals of Zootechnology, 47*, 321–333.

Bailey, D.W. & Provenza, F.D., (2008). Mechanisms determining large-herbivore distribution. In *Resource ecology* (pp. 7–28). Springer, Dordrecht.

Bardgett, R. D., & McAlister, E. (1999). The measurement of soil fungal: Bacterial biomass ratios as an indicator of ecosystem self-regulation in temperate meadow grasslands. *Biology and Fertility of Soils, 29*, 282–290.

Barnes, M. K., Norton, B. E., Maeno, M., & Malechek, J. C. (2008). Paddock size and stocking density affect spatial heterogeneity of grazing. *Rangeland Ecology & Management, 61*, 380–388.

Barrow, J. R., Lucer, M. E., Reyes-Vera, I., & Havstad, K. (2008). Do symbiotic microbes have a role in plant evolution, performance and response to stress? *Communicative and Integrative Biology, 1*, 69–93.

Blouin, M., Hodson, M. E., Delgado, E. A., Bakerd, G., et al. (2013). A review of earthworm impact on soil function and ecosystem services. *European Journal of Soil Science, 64*, 161–182.

Brain, C. K., & Sillen, A. (1988). Evidence from the Swartkrans cave for the earliest use of fire. *Nature, 336*, 64–466.

Briske, D., Derner, J., Brown, J., Fuhlendorf, S., et al. (2008). Benefits of rotational grazing on rangelands: an evaluation of the experimental evidence. *Rangeland Ecology & Management, 61*, 3–17. https://doi.org/10.2111/06-159R.1

Bryant, J. P., Provenza, F. D., Pastor, J., Reichardt, P. B., Clausen, T. P., & DuToit, J. T. (1991). Interactions between woody plants and browsing mammals mediated by secondary metabolites. *Annual Review of Ecological Systems, 22*, 431–446.

Bullock, J. M., Hill, B. C., Dale, M. P., & Silvertown, J. (1994). An experimental study of vegetation change due to sheep grazing in a species-poor grassland and the role of seedling recruitment into gaps. *Journal of Applied Ecology, 31*, 493–507.

Burkart, M. R., & Stoner, J. D. (2002). Nitrate in aquifers beneath agricultural systems. *Water Science and Technology, 45*, 19–28.

Capper, J. L. (2012). Is the grass always greener? Comparing the environmental impact of conventional, natural and grass-fed beef production systems. *Animals, 2*, 127–143.

Capper, J. L., & Bauman, D. E. (2013). The role of productivity in improving the environmental sustainability of ruminant production systems. *Annual Review of Animals and Bioscience, 1*, 469–489.

Coley, P. D., Bryant, J. P., & Chapin, F. S. (1985). Resource availability and plant antiherbivore defense. *Science, 230*, 895–899.

Conant, R. T., Paustian, K., & Elliott, E. T. (2001). Grassland management and conversion into grasslands: effects on soil carbon. *Ecological Applications, 11*, 343–355. https://doi.org/10.2307/3060893

Council for Agricultural Science and Technology (CAST). (1999). *Animal agriculture and global food supply* (Task force report 135). CAST.

De Vries, F. T., Bloem, J., et al. (2012). Extensive management promotes plant and microbial nitrogen retention in temperate grassland. *PLoS One, 7*, 51201.

De Vries, F. T., Thébault, E., Liirie, M., Birkhofer, K., et al. (2013). Soil food web properties explain ecosystem services across European land use systems. *PNAS, 110*, 14296–14301.

Delgado, J. A., Groffman, P. M., Nearing, M. A., Goddard, T., et al. (2011). Conservation practices to mitigate and adapt to climate change. *Journal of Soil and Water Conservation, 66*, 118A–129A.

DeRamus, H. A., Clement, T. C., Giampola, D. D., & Dickison, P. C. (2003). Methane emissions of beef cattle on forages: Efficiency of grazing management systems. *Journal of Environmental Quality, 32*, 269–277.

Dowhower, S. L., Teague, W. R., Casey, K. D., & Daniel, R. (2019). Soil greenhouse gas emissions as impacted by soil moisture and temperature under continuous and holistic planned grazing in native tallgrass prairie. *Agriculture, Ecosystems & Environment, 287*, p.106647.

Earl, J. M., & Jones, C. E. (1996). The need for a new approach to grazing management-is cell grazing the answer? *Rangeland Journal, 18*, 327–350.

Eshel, G., Shepon, A., Makov, T., & Milo, R. (2014). Land, irrigation water, greenhouse gas, and reactive 179 nitrogen burdens of meat, eggs, and dairy production in the United States. *PNAS, 111*, 11996–12001.
Ferguson, P. R., & Veizer, J. (2007). Coupling of water and carbon fluxes via the terrestrial biosphere and its significance to the Earth's climate system. *Journal of Geophysical Research, 112*, D24S06.
Ferguson, B. G., Diemont, S. A. W., Alfaro-Arguello, R., Martin, J. F., Nahed-Toral, J., Álvarez-Solís, D., et al. (2013). Sustainability of holistic and conventional cattle ranching in the seasonally dry tropics of Chiapas, Mexico. *Agric. Syst. 120*, 38–48. https://doi.org/10.1016/j.agsy.2013.05.005
Frank, D. A., & Groffman, P. A. (1998). Ungulate vs. landscape control of soil C and N processes in grasslands of Yellowstone National Park. *Ecology, 79*, 2229–2241.
Frank, D. A., McNaughton, S. J., & Tracy, B. F. (2002). The ecology of the Earth's grazing ecosystems. *Bioscience, 48*, 513–521.
Fuls, E. R. (1992). Semi-arid and arid rangelands: A resource under siege due to patch selective grazing. *Journal of Arid Environments, 22*, 191–193.
Gaspar, P., Mesías, F. J., Escribano, M., & Pulido, F. (2009). Sustainability in Spanish extensive farms (Dehesas): An economic and management indicator-based evaluation. *Rangeland Ecology and Management, 62*, 153–162.
Gattinger, A., Muller, A., Haeni, M., Skinner, C., et al. (2012). Enhanced top soil carbon stocks under organic farming. *Proceedings of the National Academy of Sciences, 109*, 18226–18231.
Gelfand, I., Cui, M. D., Tang, J. W., & Robertson, G. P. (2015). Short-term drought response of N2O and CO2 emissions from Mesic agricultural soils in the US Midwest. *Agriculture, Ecosystems and Environment, 212*, 127–133.
Gillingham, A. G., & Thorrold, B. S. (2000). A review of New Zealand research measuring phosphorus in runoff from pastures. *Journal of Environmental Quality, 29*, 88–96.
Green, L. E., Porras-Alfaro, A., & Sinsabaug, R. L. (2008). Translocation of nitrogen and carbon integrates biotic crust and grass production in desert grassland. *Journal of Ecology, 96*, 1076–1085.
Hamilton, E. W., & Frank, D. A. (2001). Can plants stimulate soil microbes and their own nutrient supply? Evidence from a grazing tolerant grass. *Ecology, 82*, 2397–2402.
Havstad, K. M., Peters, D. P. C., Skaggs, R., Brown, J., Bestelmeyer, B., Fredrickson, E., et al. (2007). Ecological services to and from rangelands of the United States. *Ecological Economics 64*, 261–268. https://doi.org/10.1016/j.ecolecon.2007.08.005
Herrick, J. E., & Lal, R. (1995). Soil physical property changes during dung decomposition in a tropical pasture. *Soil Science Society of America Journal 59*, 908–912. https://doi.org/10.2136/sssaj1995.03615995005900030040x
Holland, E. A., Parton, W. J., Detling, J. K., & Coppock, D. L. (1992). Physiological responses of plant populations to herbivory and other consequences of ecosystem nutrient flow. *American Naturalist, 140*, 685–706.
Hristov, A. N. (2012). Historic, pre-European settlement, and present-day contribution of wild ruminants to enteric methane emissions in the United States. *Journal of Applied and Natural Science, 90*, 1371–1375.
Huffaker, R., & Cooper, K. (1995). Plant succession as a natural range restoration factor in private livestock enterprises. *American Journal of Agricultural Economics, 77*, 901–913.
Jacobo, E. J., Rodríguez, A. M., Bartoloni, N., & Deregibus, V. A. (2006). Rotational grazing effects on rangeland vegetation at a farm scale. *Rangeland Ecology & Management, 59*, 249–257. https://doi.org/10.2111/05-129R1.1
Jakoby, O., Quaas, M. F., Müller, B., Baumgärtner, S., & Frank, K. (2014). How do individual farmers' objectives influence the evaluation of rangeland management strategies under a variable climate? *J. Appl. Ecol. 51*, 483–493. https://doi.org/10.1111/1365-2664.12216

Jakoby, O., Quaas, M. F., Baumgärtner, S., & Frank, K. (2015). Adapting livestock management to spatio-temporal heterogeneity in semi-arid rangelands. *Journal of Environmental Management, 162*, 179–189.
James, E., Kleinman, P., Veith, T., Stedman, R., et al. (2007). Phosphorus contributions from pastured dairy cattle to streams of the Cannonsville Watershed, New York. *Journal of Soil and Water Conservation, 62*, 40–47.
Jandl, R., Rodeghiero, M., Martinez, C., Cotrufo, M. F., et al. (2014). Current status, uncertainty and future needs in soil organic carbon monitoring. *Science of the Total Environment, 468*, 376–383.
Janzen, H. H. (2010). What place for livestock on a re-greening earth? *Animal Feed Science and Technology, 166–167*, 783–796.
Kaschuk, G., Kuyper, T. W., Leffelaar, P. A., Hungria, M., et al. (2009). Are the rates of photosynthesis stimulated by the carbon sink strength of rhizobial and arbuscular mycorrhizal symbioses? *Soil Biology and Biochemistry, 41*, 1233–1244.
Khidir, H. H., Porras-Alfaro, A., Herrera, J., et al. (2008). A general suite of fungal endophytes dominate the roots of two dominant grasses in a semiarid grassland. *Journal of Arid Environments, 74*, 35–42.
Knapp, A. K. (1985). Effect of fire and drought on the ecophysiology of Andropogon gerardii and Panicum virgatum in a tallgrass prairie. *Ecology, 66*, 1309–1320.
Kobayashi, M., Howitt, R. E., Jarvis, L. S., & Emilio, A. L. (2007). Stochastic rangeland use under capital constraints. *American Journal of Agricultural Economics, 89*, 805–817.
Liebig, M. A., Gross, J. R., Kronberg, S. L., Phillips, R. L., & Hanson, J. D. (2010). Grazing management contributions to net global warming potential: A long-term evaluation in the northern Great Plains. *Journal of Environmental Quality, 39*, 799–809.
Liu, H., Prugnolle, F., Manica, A., & Balloux, F. (2006). A geographically explicit genetic model of worldwide human-settlement history. *American Journal of Human Genetics, 79*, 230–237.
López- Ridaura, S., Masera, O., & Astier, M. (2002). Evaluating the sustainability of complex socio-environmental systems: The MESMIS framework. *Ecological Indicators, 2*, 135–148.
Machmuller, M. B., Kramer, M. G., Cyle, T. K., Hill, N., et al. (2015). Emerging land use practices rapidly increase soil organic matter. *Nature Communications, 6*, 6995. https://doi.org/10.1038/ncomms7995
Martin, R., Müller, B., Linstädter, A., & Frank, K. (2014). How much climate change can pastoral livelihoods tolerate? Modelling rangeland use and evaluating risk. *Global Environmental Change, 24*, 183–192.
Massy, C. (2018). *Call of the reed warbler: A new agriculture, A New Earth*. Chelsea Green Publishing.
Mathews, B. W., Sollenberger, L. E., Nair, V. D., & Staples, C. R. (1994). Impact of grazing on soil nitrogen, phosphorus, potassium, and sulfur distribution. *Journal of Environmental Quality, 23*, 1006–1013.
McNaughton, S. J., Oesterheld, M., Frank, D. A., & Williams, K. J. (1989). Ecosystem-level patterns of primary productivity and herbivory in terrestrial habitats. *Nature, 341*, 142–144.
Millennium Ecosystem Assessment (MEA). (2005). *Ecosystems and human well-being: Synthesis*. Island Press.
Montgomery, D. R. (2012). *Dirt: The erosion of civilizations*. University of California Press.
Morriën, E., Hannula, S. E., Snoek, L. B., Helmsing, N. R., et al. (2017). Soil networks become more connected and take up more carbon as nature restoration progresses. *Nature Communications, 8*, 14349.
Mosier, S. (2020). *Mechanisms and management for soil carbon sequestration*. PhD. thesis, Colorado State University, Fort Collins, Co.
Mottet, A., de Haan, C., Falcucci, A., Tempio, G., et al. (2017). Livestock: On our plates or eating at our table? A new analysis of the feed/food debate. *Global Food Security, 14*, 1–8.
Müller, B., Frank, K., & Wissel, C. (2007). Relevance of rest periods in non-equilibrium rangeland systems: A modelling analysis. *Agricultural Systems, 92*, 295–317.

Müller, B., Schulze, J., Kreuer, D., Linstädter, A., & Frank, K. (2014). How to avoid unsustainable side effects of managing climate risk in drylands – The supplementary feeding controversy. *Agricultural Systems, 139*, 153–165.

Ngumbi, E., & Kloepper, J. (2016). Bacterial-mediated drought tolerance: Current and future prospects. *Applied Soil Ecology, 105*, 109–125.

Nielsen, U. N., Ayres, E., Wall, D. H., & Bardgett, R. D. (2011). Soil biodiversity and carbon cycling: A review and synthesis of studies examining diversity-function relationships. *European Journal of Soil Science, 62*, 105–116.

Norton, B. E. (1998). The application of grazing management to increase sustainable livestock production. *Animal Production Science, 22*, 15–26.

Norton, B. E. (2003). Spatial management of grazing to enhance livestock production and resource condition: A scientific argument. In N. Allsopp, A. R. Palmer, S. J. Milton, K. P. Kirkman, G. I. H. Kerley, C. R. Hurt, & C. J. Brown (Eds.), *Proceedings of the 7th International Rangeland Congress, Durban, South Africa* (pp. 810–820). Document Transfer Technologies, Irene, Republic of South Africa.

O'Connor, T. G. (1992). Patterns of plant selection by grazing cattle in two Savanna grasslands: A plant's eye view. *Journal of the Grassland Society of Southern Africa, 9*, 97–104.

Odum, E. P. (1997). *Ecology: A bridge between science and society*. Sinauer.

Ottoboni, A., & Ottoboni, F. (2013). *The modern nutritional diseases and how to prevent them* (2nd ed.). Vicente Books.

Park, Y., Ale, S., Teague, W. R., & Jeong, J. (2017). Evaluating the ranch and watershed scale impacts of using traditional and adaptive multi-paddock grazing on runoff, sediment, and nutrient losses in North Texas. *Agriculture, Ecosystems and Environment, 240*, 32–44.

Parsons, S. D., & Pratt, D. W. (1991). *Ranching for profit-self-study course*. Ranch Management Consultants Inc.

Pokorný, J., Brom, J., Čermák, J., Hesslerová, P., Huryna, H., Nadezhdina, N., & Rejšková, A. (2010). Solar energy dissipation and temperature control by water and plants. *International Journal of Water, 5*, 311–336.

Provenza, F. D. (2008). What does it mean to be locally adapted and who cares anyway? *Journal of Animal Science, 86*, 271–284. https://doi.org/10.2527/jas.2007-0468

Provenza, F. D., Villalba, J. J., Dziba, L. E., Atwood, S. B., & Banner, R. E. (2003a). Linking herbivore experience, varied diets, and plant biochemical diversity. *Small Ruminant Research, 49*, 257–274.

Provenza, F. D., Villalba, J. J., & Bryant, J. P. (2003b). Foraging by herbivores, linking the biochemical diversity of plants with herbivore culture and landscape diversity. In J. A. Bissonette & I. Storch (Eds.), *Landscape ecology and resource management* (Linking theory with practice) (pp. 387–421). Island Press.

Ragab, R., & Prudhomme, C. (2002). Soil and water: Climate change and water resources management in arid and semiarid regions: Prospective and challenges for the 21st century. *Biosystems Engineering, 81*, 3–34.

Retallack, G. J. (2013). Global cooling by grassland soils of the geological past and near future. *Annual Review of Earth and Planetary Sciences, 41*, 69–86.

Richardson, P. Q., & Richardson, R. H. (2000). Dung beetles improve the soil community in Texas and Oklahoma. *Ecol. Rest. 18*, 116–117.

Rillig, M. C. (2004). Arbuscular mycorrhizae and terrestrial ecosystem processes. *Ecol. Lett. 7*, 740–754. https://doi.org/10.1111/j.1461-0248.2004.00620.x

Ripple, W. J., Smith, P., Haberl, H., Montzka, S. A., et al. (2014). Ruminants, climate change and climate policy. *Nature Climate Change, 4*, 2–5.

Ritten, J. P., Frasier, W. M., Bastian, C. T., & Gray, S. T. (2010). Optimal rangeland stocking decisions under stochastic and climate impacted weather. *American Journal of Agricultural Economics, 92*, 1242–1255.

Ritter, W. F. (1988). Reducing impacts of non-point source pollution from agriculture. *Journal of Environmental Science and Health, 23*, 645–667.

Rodgers, D., & Wolf, R. (2020). *Sacred cow: The case for (better) meat: Why well-raised meat is good for you and good for the planet* (p. 304). BenBella Books.
Rodriguez, H., & Fraga, R. (1999). Phosphate solubilizing bacteria and their role in plant growth promotion. *Biotechnology Advances, 17*, 319–339.
Rowntree, J. E., Ryals, R., DeLonge, M. S., Teague, W. R., et al. (2016). Potential mitigation of Midwest grass-finished beef production emissions with soil carbon sequestration in the United States of America. *Future of Food: Journal on Food, Agriculture & Society, 4*, 31–38.
Ruser, R., Flessa, H., Russow, R., Schmidt, G., Buegger, F., & Munch, J. C. (2006). Emission of N2O, N2 and CO2 from soil fertilized with nitrate: Effect of compaction, soil moisture and rewetting. *Soil Biology and Biochemistry, 38*, 263–274.
Saggar, S., Hedley, C. B., Giltrap, D. L., & Lambie, S. M. (2007). Measured and modelled estimates of nitrous oxide emission and methane consumption from a sheep-grazed pasture. *Agriculture, Ecosystems and Environment, 122*, 357–365.
Sauer, T. J., Daniel, T. C., Nichols, D. J., West, C. P., et al. (2000). Runoff quality from poultry litter-treated pasture and forest sites. *Journal of Environmental Quality, 29*, 515–521.
Savory, A., & Butterfield, J. (2016). *Holistic management* (3rd ed.). Island Press.
Sharpley, A. N., Chapra, S. C., Wedepohl, R., Sims, J. T., et al. (1994). Managing agricultural phosphorus for protection of surface waters: Issues and options. *Journal of Environmental Quality, 23*, 437–451.
Slade, E. M., Riutta, T., Roslin, T., & Tuomisto, H. L. (2016). The role of dung beetles in reducing greenhouse gas emissions from cattle farming. *Nature Scientific Reports, 6*, 18140.
Smith, K. A., Ball, T., Conen, F., Dobbie, K. E., Massheder, J., & Rey, A. (2003). Exchange of greenhouse gases between soil and atmosphere: Interactions of soil physical factors and biological processes. *European Journal of Soil Science, 54*, 779–791.
Sovell, L. A., Vondracek, B., Frost, J. A., & Mumford, K. G. (2000). Impacts of rotational grazing and riparian buffers on physio-chemical and biological characteristics of Southeastern Minnesota, USA, streams. *Environmental Management, 26*, 629–641.
Stackhouse-Lawson, K.R., Rotz, C.A., Oltjen, J.W., & Mitloehner, F.M. (2012). Carbon footprint and ammonia emissions of California beef production systems. *Journal of Animal Science, 90*(12), pp.4641–4655.
Stanley, P. L., Rowntree, J. E., Beede, D. K., DeLonge, M. S., & Hamm, M. W. (2018). Impacts of soil carbon sequestration on life cycle greenhouse gas emissions in Midwestern USA beef finishing systems. *Agricultural Systems, 162*, 249–258.
Stout, W. L., Weaver, S. R., Gburek, W. J., Folmar, G. J., et al. (2000). Water quality implications of dairy slurry applied to cut pastures in the Northeast USA. *Soil Use and Management, 16*, 189–193.
Teague, R., & Barnes, M. (2017). Grazing management that regenerates ecosystem function and grazingland livelihoods. African J*ournal of Range & Forage Science, 34*(2), 77–86.
Teague, W. R., Dowhower, S. L., & Waggoner, J. A. (2004). Drought and grazing patch dynamics under different grazing management. *Journal of Arid Environments, 58*, 97–117.
Teague, W. R., Dowhower, S. L., Baker, S. A., Haile, N., DeLaune, P. B., & Conover, D. M. (2011). Grazing management impacts on vegetation, soil biota and soil chemical, physical and hydrological properties in tall grass prairie. *Agriculture, Ecosystems & Environment. 141*, 310–322. https://doi.org/10.1016/j.agee.2011.03.009
Teague, R., Provenza, F., Kreuter, U., Steffens, T., & Barnes, M. (2013). Multi-paddock grazing on rangelands: Why the perceptual dichotomy between research results and rancher experience? *Journal of Environmental Management, 128*, 699–717.
Teague, R., Grant, W., & Wang, H. (2015). Assessing optimal configurations of multi-paddock grazing strategies in tallgrass prairie using a simulation model. *Journal of Environmental Management, 150*, 262–273.
Teague, W. R., Apfelbaum, S., Lal, R., Kreuter, U. P., Rowntree, J., Davies, C. A., Conser, R., Rasmussen, M., Hatfield, J., Wang, T., Wang, F., & Byck, P. (2016). The role of ruminants

in reducing agriculture's carbon footprint in North America. *Journal of Soil and Water Conservation, 71*, 156–164.

Thurow, T. (1991). Hydrology and erosion. In R. K. Heitschmidt & J. W. Stuth (Eds.), *Grazing management: An ecological perspective* (pp. 141–159). Timberland Press.

U.S. Department of Agriculture Economic Research Service, using National Agricultural Statistics Service data. (2021). *Cattle & Beef, Sector at a Glance*. https://www.ers.usda.gov/topics/animal-products/cattle-beef/sector-at-a-glance/

Vadas, P. A., Busch, D. L., Powell, J. M., & Brink, G. E. (2014). Monitoring runoff from cattle-grazed pastures for a phosphorus loss quantification tool. *Agriculture, Ecosystems and Environment, 199*, 124–131.

Van der Heijden, M. G. A., Bardgett, R. D., & Van Straalen, N. M. (2008). The unseen majority: Soil microbes as drivers of plant diversity and productivity in terrestrial ecosystems. *Ecological Letters, 11*, 296–310.

Veizer, J., Godderis, Y., & Francois, L. M. (2000). Evidence for decoupling of atmospheric CO2 and global climate during the Phanerozoic eon. *Nature, 408*, 689–701.

Walker, B. H., Carpenter, S., Anderies, J., Abel, N., Cumming, G., Janssen, M., Lebel, L., Norberg, J., Peterson, G. D., & Pritchard, R. (2002). Resilience management in social-ecological systems: A working hypothesis for a participatory approach. *Conservation Ecology, 6*, 14.

Wang, C., Han, G., Wang, S., Zhai, X., et al. (2014). Sound management may sequester methane in grazed rangeland ecosystems. *Scientific Reports*. https://doi.org/10.1038/srep04444

Wang, T., Teague, W. R., & Park, S. C. (2016). Evaluation of continuous and multipaddock grazing on vegetation and livestock performance - a modeling approach. *Rangeland Ecology & Management, 69*, 457–464. https://doi.org/10.1016/j.rama.2016.07.003

Wang, T., Teague, W. R., Park, S. C., & Bevers, S. (2015). GHG mitigation potential of different grazing strategies in the United States Southern Great Plains. *Sustainability, 7*, 13500–13521.

Wardle, D. A., & Bardgett, R. D. (2004). Human-induced changes in large herbivorous mammal density: the consequences for decomposers. *Frontiers in Ecology and the Environment, 2*(3), 145–153.

Webber, D. F., Mickelson, S. K., Ahmed, S. I., Russell, J. R., et al. (2010). Livestock grazing and vegetative filter strip buffer effects on runoff sediment, nitrate, and phosphorus losses. *Journal of Soil and Water Conservation, 65*, 34–41.

Weltz, M. A., Jolley, L., Goodrich, D., Boykin, K., et al. (2011). Techniques for assessing the environmental outcomes of conservation practices applied to rangeland watersheds. *Journal of Soil and Water Conservation, 66*, 154A–162A.

West, P. C., Gerber, J. S., Engstrom, P. M., Muelle, N. D., et al. (2014). Leverage points for improving global food security and the environment. *Science, 345*, 325–328.

Chapter 7
The Limits of Grass

Diane Mayerfeld

Abstract As the previous chapter lays out, where livestock is raised with regenerative grazing management the net greenhouse gas intensity can be lower than that for industrial production, if soil carbon fluxes are taken into account. And livestock don't just provide income and food; their manure, hoof action, and periodic grazing appear to play a role in helping soils store carbon. However, we cannot assume that the extraordinary gains in soil carbon that a few farms have seen can be replicated everywhere and sustained indefinitely, especially in the face of warming global temperatures. We also need to acknowledge that there are trade-offs between managing grazing systems to maximize soil carbon and maximizing milk or meat production. Finally, we need far more research on how specific grazing management practices can affect both enteric emissions and soil carbon storage.

Keywords Climate change · Grazing · Greenhouse gas emissions · Livestock · Soil carbon

7.1 Introduction

As Teague and Apfelbaum lay out in the previous chapter, where livestock is raised with regenerative grazing management the net greenhouse gas intensity can be significantly lower than that for industrial production, if soil carbon fluxes are taken into account. In some cases grass-based livestock systems may even act as a net sink for greenhouse gases, because the atmospheric CO_2 that the pasture or range converts to carbon stored in the soil exceeds the CO_2eq of the methane and nitrous oxide emissions from the livestock.[1] And the livestock don't just provide income

[1] See also (General Mills, n.d.; Rowntree et al., 2020; Stanley et al., 2018)

D. Mayerfeld (✉)
Division of Extension, Agriculture Institute, University of Wisconsin, Madison, WI, USA
e-mail: dbmayerfeld@wisc.edu

D. Mayerfeld (ed.), *Our Carbon Hoofprint*, Food and Health,
https://doi.org/10.1007/978-3-031-09023-3_7

and food; their manure and hoof action and periodic grazing appear to play a key role in maintaining a diverse and thriving plant and soil biology, and so helping soils store carbon.[2] However, we need to be cautious about assuming that the extraordinary gains in soil carbon that some farms have seen can be replicated everywhere and sustained indefinitely, especially in the face of warming global temperatures. We also need to acknowledge some limits of grass-based livestock production, and the need for more research on how specific grazing management practices affect both enteric emissions and soil carbon storage.

First, let me return to the need for good information and honest negotiation to tackle the climate crisis. In 2013 the well-known advocate of rotational grazing and holistic farm management, Allan Savory, gave a TED talk in which he claimed that properly managed rotational grazing could not only reverse desertification, but that it could also reduce atmospheric CO_2 to pre-industrial levels.[3] Over the past 7 years the talk has had more than 7 million views, and generated both excitement and consternation in the grazing community. The idea that improving how we raise livestock is an important part of stabilizing our climate, and that better grazing management is critical is the central argument of this book. However, to suggest that better grazing management alone can bring our atmospheric carbon, methane, and nitrous oxide down to pre-industrial levels is deeply irresponsible. In the best possible scenario, with full support from all levels of government, civil society, and business, better grazing management could theoretically restore the carbon content of grassland soils to pre-industrial levels in a human lifetime. But even that herculean achievement would not compensate for the carbon dioxide and methane emitted by fossil fuel production and combustion, the methane emitted from warming permafrost soils, or the many other sources of greenhouse gases humans have unleashed.[4]

The claim that better grazing management alone can reverse climate change is just as false and irresponsible as the claim that eliminating meat from the diet can solve the climate crisis. It may excite grazing enthusiasts, but it also tends to discredit more realistic arguments for grazing. Worse yet, like the exaggerated claims for a vegan diet, it suggests that we do not need to rein in emissions from transportation, electricity, heating, and modern levels of consumption in general.

The extraordinary increases in soil carbon that some grazing advocates have achieved are probably not feasible everywhere. And just as proponents of industrial livestock production are wrong to ignore the soil carbon implications of perennial pasture or range versus harvested annual feed crops such as maize and soybeans, grazing advocates cannot focus solely on soil carbon and ignore methane and nitrous oxide emissions from grazing systems.

---

[2] Katz-Rosene (2020), Frith (2020).

[3] Savory (2013).

[4] Fargione et al. (2018), Schlesinger and Amundson (2019) Theoretically carbon storage in soils in the U.S. could offset 21% of the country's greenhouse gas emissions, or worldwide efforts could offset 5% of emissions, and there are significant social and economic barriers to achieving even those levels of carbon storage.

7.2 Methane and Ruminants

There is an argument that methane, the primary greenhouse gas emitted as a result of beef and dairy production, should not be of much concern because its average residence time in the atmosphere is only about 12 years. This view suggests that because of its relatively short duration methane won't build up in the atmosphere, and so we should instead focus our emission reduction efforts on CO_2, which can remain in the atmosphere for thousands of years.[5] However, atmospheric methane concentrations have more than doubled since 1750, from about 700 ppb to 1857 ppb in 2018. In fact, proportionately atmospheric methane concentrations have increased more than CO_2, though the warming contribution of CO_2 has been far greater because the total quantity is so much larger.[6] In the U.S. enteric emissions account for roughly 23% of all anthropogenic methane emissions, so they are not insignificant.[7]

Moreover, the CO_2 equivalence measures used in this book and in most climate change discussions already take the longevity of greenhouse gases into account, although there is debate about how to weigh the atmospheric lifespan of different gases. Table 7.1 shows estimates of the warming potential of key greenhouse gases over both a 20-year period and a 100-year period using two different measures: Global Warming Potential (GWP) and Global Temperature change Potential (GTP). The measure used in this book and in most climate change literature is the 100-year Global Warming Potential. As our understanding of greenhouse gases grows, these calculations continue to be adjusted. Thus, in the second IPCC (United Nations Intergovernmental Panel on Climate Change) assessment, the 100-year GWP for methane was thought to be 21; when *Livestock's Long Shadow* was published in 2006 they used a 100-year GWP for methane of 23; and since 2014 the IPCC assessment's 100-year GWP for methane has been calculated as 28.[8]

Table 7.1 CO_2eq values for methane (CH_4) and nitrous oxide (N_2O)

		GWP		GTP	
Greenhouse gas	Lifetime (yrs)	Cumulative forcing over 20 years	Cumulative forcing over 100 years	Temperature change after 20 years	Temperature change after 100 years
CO_2	[a]	1	1	1	1
CH_4	12.4	84	28	67	4
N_2O	121	264	265	277	234

IPCC (2014), p. 87

[a]No single lifetime can be given for CO_2. GWP stands for Global Warming Potential; GTP for Global Temperature change Potential

[5] U.S. Environmental Protection Agency (2020), Cain et al. (2019), Mitloehner (2020).

[6] U.S. Environmental Protection Agency (2020) page 1–4

[7] Maasakkers et al. (2016).

[8] "Global Warming Potential Values" (n.d.).

The question of how to compare the impact of different greenhouse gases is difficult. There is no single right answer. It is important to understand and think about the impacts over many different time frames: in the next 10 or 20 years, over a human lifetime (roughly 100 years), and over multiple generations (500 years or longer). The world is experiencing severe climate impacts right now, and reducing methane emissions can help mitigate those impacts in the short term. On the other hand, if we fail to bring down our CO_2 emissions because we rely too much on the short-term climate benefits of reducing methane, then our children and grandchildren will suffer even more horrendous climate change. At the same time, comparing the impacts of different greenhouse gases is confusing even when a single timeframe is used, which is why the accepted standard time frame of 100 years is a useful compromise for most discussions.

The fact that methane is a relatively short-lived greenhouse gas does not give animal agriculture license to ignore it. Increases in methane emissions due to growing livestock numbers have very significant climate impacts, and conversely, reductions in methane emissions from livestock could have immediate climate benefits. For methane as for CO_2, moving away from our dependence on fossil fuels to sustainable renewable energy sources offers the greatest short- and long-term climate benefits for the U.S. as a whole.[9] However, methane emissions from enteric fermentation and manure storage account for more than half of animal agriculture's greenhouse gas emissions, even in the U.S.[10]

The previous chapter mentions emerging research indicating that in some cases grazing livestock may have lower methane emissions from ruminant digestion than cattle fed a standard industrial diet. However, most research to date has found that cows fed diets that include grain and added sugar emit less methane than cattle feeding only on pasture. In general, the less the animal has to rely on anaerobic microbes in the rumen to break down cellulose and other hard to digest materials, the less methane is generated.[11] For this reason, cattle in concentrated animal feeding operations (CAFOs) tend to have lower enteric methane emissions per unit of meat or milk produced than cattle on pasture or range.[12] However, the composition and management of the forage in the pasture or range also matter: high quality pasture and

[9] The increase in atmospheric methane that has occurred over the past 250 years is not only or even primarily due to animal agriculture. In the U.S. the main source of anthropogenic methane emissions is fossil fuel production, processing, and distribution. The combination of enteric emissions and manure emissions from animal agriculture are the next biggest human-caused source of methane in the U.S., followed by emissions from waste disposal, primarily landfills. (Maasakkers et al., 2016) Global methane emissions from coal, oil, and natural gas production are growing faster than methane emissions from livestock and are estimated to be nearly equal to livestock emissions in Chapter 5 of the preliminary IPCC report released in August 2021, which is subject to revisions following the "Summary for Policy Makers" approval, corrigenda, copy-editing, and layout (IPCC, 2021).

[10] U.S. Environmental Protection Agency (2020), U.S. Department of Agriculture (2016a, b).

[11] Beauchemin et al. (2020).

[12] Herrero et al. (2016).

range result in lower enteric methane emissions per unit of milk or meat than low quality pasture.[13]

Manure methane emissions are generally lower for grazing animals than for livestock kept indoors. The manure from dairy cows and pigs in efficient industrial systems is typically stored in anaerobic conditions, which are a perfect environment for microbes that release methane. On the other hand, if industrial livestock operations use manure digesters to capture the methane from the manure and use that captured methane to replace fossil fuels, then their manure methane emissions are negligeable. Indeed, some models then consider those farms to be greenhouse gas sinks rather than sources, because the captured methane can replace fossil fuels that would otherwise have generated significant greenhouse gas emissions in their production and combustion.[14] Without subsidies, however, building and operating manure digesters can be costly, and in the U.S. at present most methane from stored manure is escaping to the atmosphere.[15]

As we compare methane emissions from grazing operations to those from concentrated industrial farms, once again we have a situation that is complex and not fully understood. Overall, enteric emission intensity is likely higher from grazing livestock, both because there is a lot of variability in the quality of the forage available in pasture and range systems, and because growth and milk production are usually slower in grazing systems than in concentrated animal feeding operations (CAFOs). On the other hand, manure methane emissions are usually lower in grazing systems than those from CAFOs. However, when industrial livestock operations use manure digesters their manure methane emissions are as low or lower than those from grazing farms. At the same time, there is a wide range in methane emission intensity from grazing operations, depending on the quality of the forage and the genetics of the grazing animal, and we still have a lot to learn about optimum forage and grazing management to minimize enteric emissions. Despite this complexity and uncertainty, there are promising pathways to significantly reduce methane emissions from both industrial and grazing livestock systems. Industrial systems can work on improving the economic viability and management of manure digestion and on feed additives that reduce enteric emissions, while grazing systems can seek to optimize the quality of the forage on the pasture and range to reduce enteric emissions from grazing livestock. In addition, there are opportunities for industrial

[13] Zubieta et al. (2021), Beauchemin et al. (2020), Dini et al. (2012), Rowntree et al. (2016).

[14] See for example (Aguirre-Villegas et al. 2015, b). However, this accounting is misleading. It is true that digesters convert the final greenhouse gas emission from methane to CO_2, which has a lower global warming potential. It is also true that capturing and burning the methane to create heat, electricity, or transportation fuel reduces the need for fossil fuels, with their enormous climate impacts. However, the farm still does not act as a net carbon sink in the way land use change to forest or grassland would. Even with a manure digester the farm is still a net greenhouse gas emitter unless it stores more carbon in the soil than the sum of its enteric, nitrous oxide, and other greenhouse gas emissions.

[15] U.S. Environmental Protection Agency (n.d.-a, n.d.-b, n.d.-c).

systems to incorporate some grazing, as well as for grazing systems to include some efficiencies of industrial livestock systems.

7.3 The Confounding Role of Nitrous Oxide

Methane emissions from livestock digestion and manure usually receive the most attention, because those are the largest source of livestock greenhouse gas emissions in countries that are not converting forest to pasture. Because manure from grazing livestock decomposes in aerobic conditions it results in minimal methane emissions (Table 7.2), and in a grazing setting those small emissions can be offset by the ability of soil organisms to oxidize methane. However, those aerobic conditions can encourage the conversion of the nitrogen contained in the manure to nitrous oxide, which is an extremely potent and relatively long-lived greenhouse gas.[16] As a result, IPCC models assume that manure and urine in grazing systems have higher N_2O emissions than manure from industrial systems, and, as Table 7.2 shows, the U.S. Department of Agriculture estimates that nitrous oxide emissions from grazed land account for more than a quarter of U.S. livestock greenhouse gas emissions. Other research suggests that practices that promote soil carbon gains, such as

Table 7.2 U.S. Greenhouse Gas Emissions by Livestock Category and Source in 2013

	Enteric fermentation	Managed livestock waste		Grazed land			
	CH_4	CH_4	N_2O	N_2O^a	CH_4	CO_2	**Total**
Animal type	*MMT CO_2 eq.*						
Beef Cattle	117.10	0.62	7.65	85.16	2.38	2.95	**215.87**
Dairy Cattle	41.59	31.66	5.74	5.06	0.11	0.18	**84.34**
Swine	2.47	23.05	1.89	0.24	0.01	0.01	**27.66**
Horses	1.59	0.02	0.12	3.44	0.21	0.12	**5.49**
Poultry	NA	3.22	1.58	0.17	0.01	0.01	**4.98**
Sheep	1.07	0.03	0.31	0.80	0.04	0.03	**2.28**
Goats	0.31	0.00	0.02	0.64	0.02	0.02	**1.02**
American Bison	0.32	NA	NA	0.32	0.01	0.01	**0.66**
Mules and Asses	0.07	0.00	0.00	0.10	0.01	0.00	**0.19**
Total	**164.53**	**58.61**	**17.3**	**95.93**	**2.78**	**3.53**	**342.49**

U.S. Department of Agriculture (2016a, b), Table 7.2–7.2
Note: Methane emissions from manure deposited on grasslands is not partitioned by animal type. MMT CO_2 eq. is million metric tons carbon dioxide equivalent. CH_4 is methane; N_2O is nitrous oxide; CO_2 is carbon dioxide
[a]Includes direct and indirect emissions

[16] Butterbach-Bahl et al. (2013).

reducing tillage, applying and incorporating manure, and incorporating crop residue, also tend to increase nitrous oxide emissions from soils.[17]

The fate of nitrogen from manure needs to be kept in mind, but there is enormous variation in N_2O emissions from pasture and rangeland, and considerably more research is needed on this question.[18] Moreover, although direct nitrous oxide emissions from manure lagoons are lower than N_2O emissions from manure on pasture, the ammonia and ammonium generated from anaerobic manure storage systems can later be transformed to nitrous oxide in soils, and under certain conditions recommended management practices like manure injection on cropland or reducing tillage can increase emissions.[19] Table 7.2 is also misleading because it does not include the emissions from crops raised to feed livestock. The U.S. Department of Agriculture attributes those emissions to crop production rather than to livestock, but since at least half the cropland in the U.S. produces livestock feed, a portion of the estimated 168 MMT CO_2eq nitrous oxide emissions from cropland should also be allocated to livestock production.[20]

A first step to reducing direct N_2O emissions from all livestock is to avoid feeding them excess protein.[21] Adequate protein levels in the feed are needed for healthy animal growth and milk production, but excess dietary protein does not have health benefits and increases the risk of nitrous oxide emissions from the manure.[22] While farmers know that a sudden increase in protein content in ruminant diets is dangerous for the animals, in general the perception is that as long as the animals do not suffer bloat, diets high in protein are good for production. The environmental concerns associated with high nitrogen levels in manure have not received much attention in outreach to farmers or in the discussions around livestock's contributions to climate change.

[17] Li et al. (2005).

[18] Dangal et al. (2019), Jackson et al. (2015), Nichols et al. (2016), Wolf et al. (2010).

[19] Adair et al. (2018), VanderZaag et al. (2011).

[20] It is surprisingly difficult to determine what portion of U.S. cropland is used to raise livestock feed. The U.S. Department of Agriculture estimates that maize (corn) accounts for roughly 95% of livestock feedgrain. However, this statistic does not account for the soybean meal used for livestock feed after the oil is extracted for food and industrial use, and it does not count non-grain feed crops such as alfalfa or corn silage. Within corn grown for grain it is also challenging to estimate what portion of greenhouse gas emissions to attribute to livestock. In 2020 the U.S. Department of Agriculture estimated that 43% of the U.S. corn grain crop was used to feed livestock. In the same year 46% of the U.S. crop was used for "food, alcohol, and residual use," most of which was ethanol used as a gasoline additive. After the sugars are extracted from the maize and made into ethanol, most of the residual distillers grain is fed to livestock. So how many of the greenhouse gas emissions from soybeans and grain for ethanol should be attributed to livestock?

[21] Gerber et al. (2013), Wattiaux et al. (2019), Hristov et al. (2013), Montes et al. (2013).

[22] High dietary protein levels may not increase production well before reaching levels that threaten cow health, e.g., (Dini et al., 2012).

7.4 Winter

One limitation of grazing systems is that in most cases they have periods when the grass doesn't grow. Take my home state of Wisconsin, where about 34,000 farms have pasture.[23] Most years the grass on those pastures is beautiful and lush in May and June. Cows, sheep, and goats grazing on Wisconsin's pastures in early summer gain weight and produce plenty of milk. The grass grows faster than the animals eat it all, and farmers harvest some of their pastures for hay before the grass goes to seed and loses quality. In July grass growth usually slows down as the weather gets hotter and drier. Farmers complain about the summer slump, when it can take 45 or even 60 days for the grass in a paddock to regrow after grazing, unlike that spring flush when the grass might fully regrow in 4 weeks.[24] But good grazing farms have enough pasture and paddocks to see their herd through a normal summer slump, and in September as the weather cools the grass growth usually picks up again. Then in October, the weather grows cooler still, temperatures at night go below 0°C, and pasture growth slows down again, and by November even in southern Wisconsin the grass is no longer growing at all.

Farmers who are committed to grazing try to keep their cows on pasture as much of the year as possible. Cows can eat the grass even when it is not actively growing, so farmers can stockpile forages for the late fall and early winter; in other words, they can keep rotating their animals to paddocks that still have plenty of grass from before winter started. The quality of that stockpiled forage can be reasonably good, though not as wonderful as the grass in May and June. But by December most Wisconsin grazing farms have to provide supplementary feed to their animals. The pastures will start to grow again in April, but that still leaves 4–6 months of the year when grazing farms here have to provide some kind of stored feed for their animals.

Farmers who are committed to grazing feed their livestock hay (preserved grass) through the winter, and cows and sheep can survive the winter on just hay. However, the hay is not as nutritious as the growing grass, and milk production and growth decrease on a diet of just hay. From a greenhouse gas emissions perspective animals produce as much (or more) methane when they digest hay as when they are on fresh pasture, but because they do not produce as much milk or meat their emission intensity goes up. If they get supplemental grain, however, they can maintain good production levels through the winter without increasing their enteric or manure methane emissions.

Across northern North America, Europe, and Asia winter limits the climate value of relying entirely on grazing. In other parts of the world the limiting factor can be a dry season. Nature and traditional farming systems responded to seasonal variation in forage growth by a combination of migration and accepting loss of

[23] USDA-NASS (2019a).

[24] A paddock is a subdivision within a pasture. Farmers subdivide the pasture into many paddocks so they can move the herd around and allow the grass in areas that have been grazed to recover.

production.[25] In some cases seasonal migration pathways have been disrupted by modern state and private property boundaries. In all cases a drop in production increases the emission intensity of the meat or milk produced by the animals. Feeding grain and other supplements during these seasons can significantly reduce the overall emission intensity of meat and milk from grazing systems.

7.5 Poultry and Pigs

Cattle, sheep, and goats can live entirely on grass and other plants that humans cannot digest. But while chickens and pigs will eat some grass and clover if they have access to pasture, their digestive systems are more similar to those of humans than to ruminants. That means that even if they are on good pasture they need supplementary feed such as grains or tubers that humans could eat. There is also little research on the effects of grazing poultry or pigs on soil carbon. Because pigs like to root and essentially till the soil surface, it is unlikely that pastured pigs would promote soil carbon storage. Both in the U.S. and globally people eat more poultry than beef, and worldwide pork consumption is also higher than beef, though in the U.S. per capita beef consumption is still slightly higher than pork.[26] There is little reason to believe that moving pork and poultry production to pasture systems would lead to increases in soil carbon. On the other hand, according to the FAO, "backyard" poultry and pork production do not have higher greenhouse gas emissions than industrial systems.[27] Thus, if pasture-based poultry or pork production systems are preferred for reasons of animal welfare, as long as the supplemental feed is produced sustainably they also likely won't increase greenhouse gas emissions compared to industrial systems.[28]

7.6 The Limits of Soil Carbon

The question of how much carbon grasslands can store is also far more complicated than the successes of a few farmers have led some to believe. Many farmers who practice grazing hear that grazing systems build soil carbon without focusing on the specific management needed to get those carbon increases or on the factors that can limit soil carbon storage even under best management. In some cases there are also trade-offs between building soil carbon and other greenhouse gas emissions.

---

[25] Human-managed migratory grazing systems are called transhumance and occur around the world (Feldt, 2020; Oteros-Rozas et al., 2013; Starrs, 2018)

[26] Winders and Ransom (2019), U.S. Department of Agriculture Economic Research Service (2020, 2021).

[27] MacLeod et al. (2013).

[28] Jakobsen et al. (2015).

The term "grazing" encompasses a wide range of practices, and that range is the source of much of the disagreement over whether grass-based or industrial meat is better for the climate. *Livestock's Long Shadow* and the successor UN Food and Agriculture Organization report *Greenhouse gas emissions from ruminant supply chains* compared emissions from grazing and mixed systems[29] worldwide, and many other analyses draw on those global systems comparisons.[30] Because mixed (industrial) systems for ruminants are concentrated in Europe and North America, while the majority of grazing-only systems are located in South America, Africa, and Asia, these world-wide comparisons of grazing and mixed systems are confounded by regional influences.

Even when comparisons are done within a country or region, they typically compare a hypothetical average of grazing with a similarly artificial average for industrial systems. The few papers that acknowledge that there is variation in greenhouse gas emissions within both industrial and grazing systems find that even small management differences can affect emission intensity enough to switch which system looks better. One paper that modeled the greenhouse gas emissions from Wisconsin dairy farms found that adjusting just one management practice -- feeding grazing cows some supplemental grain -- made milk from the grazing farm slightly less emission intensive than the industrial farm.[31] Other changes, such as adjusting the time when manure is spread or changing the use of nitrogen fertilizer on pasture or on the feed crops for the industrial system would likely have similarly significant effects on the emission intensities of each system.

As mentioned above, the research that concludes that confinement systems have lower emissions typically compares contemporary well-managed confinement systems in Europe and North America to grazing systems averaged across the globe, or to U.S. grazing systems from the middle of the twentieth century.[32] In contrast, the research that concludes that grazing systems are better for the climate typically looks at farms that use adaptive multi-paddock grazing management *and that prioritize soil health* or at least environmental quality.[33] These are a small minority of grazing farms. In 2017 about 21% of U.S. farms with permanent pasture reported that they practiced rotational or managed grazing.[34] Those terms have a fairly loose

[29] Because in straight beef production the cows almost always spend their lives on grass and the calves also often graze before their final months in a feedlot, this FAO report uses the term mixed system rather than industrial or Concentrated Animal Feeding Operation (CAFO) or confined system.

[30] For example (Garnett et al., 2017)

[31] Aguirre-Villegas et al. (2017).

[32] Capper (2012), Capper et al. (2009).

[33] Frith (2020).

[34] USDA-NASS (2019a, b).

definition, and it is likely that many of those farms practice a low level of rotational grazing, such as rotating the herd at fixed intervals or among just a few paddocks.[35]

One of the most powerful arguments for the additional labor, management, and skill required for rotational or adaptive multi-paddock grazing is that this practice allows the farmer to raise more animals on the same amount of land.[36] The reasoning is that when cattle are in the same pasture continuously the animals eat all the best forages and don't eat the less palatable plants. The best forages die because they never get the chance to regrow, while eventually the plants that are less nutritious, or are thorny or toxic take over the pasture. In contrast, if the herd is forced to eat almost all the plants in a part of the pasture and then is moved to another part of the pasture, giving all the plants in the grazed area time to fully recover, then the pasture will retain the most nutritious plants and continue to support the full herd. In many cases this argument has been borne out by both farmer experience and research. Pasture and range managed rotationally also tend to perform as well or better on ecological indicators like nesting birds and insect diversity than continuously grazed land.[37]

However, when we add greenhouse gas emissions to the equation, the happy alignment between production and environmental benefit weakens. The systems that have net zero or perhaps even negative greenhouse gas emissions are those with moderate or light stocking rates, even when the pasture condition in the heavily stocked comparison continues to be good.[38] There are several reasons why high stocking emits more greenhouse gases. First, more cows on the land emit more methane per hectare. Second, high stocking rates are often supported by practices such as nitrogen fertilization, which have significant emissions of their own. And third, the efficient utilization of the forage that allows high stocking rates also means there is not as much carbon left behind to be stored in the soil.

Gabe Brown farms near Bismarck in North Dakota, and has become well-known in progressive farming circles because of his success in building soil health on his

---

[35] Just as there are many terms used to describe industrial livestock production, there are many words describing ways to manage grazing, and the distinctions between the terms are not always clear-cut. Rotational grazing means that the animals are moved between paddocks, but technically does not specify the guidelines for how soon animals need to be moved or how long recovery periods need to be. The term adaptive multi-paddock (AMP) grazing suggests that the animals are rotated depending on the condition of the pasture, but how forage condition is judged also varies depending on the forage type, livestock, and other factors. Other terms, including Holistic grazing and managed grazing similarly rely on the expert judgment of the farmer rather than on defined rules. The lack of clear prescriptions allows farmers to respond to real ecological conditions (Lyon et al., 2011), but it also relies on a high level of skill and experience on the part of the farmer, and it can lead to the false perception that any movement of livestock will provide significant ecological benefits.

[36] Briske et al. (2011), Franzluebbers et al. (2012), Oates et al. (2011), Teague et al. (2013), Kriegl and McNair (2005).

[37] Spratt et al. (2021), Franzluebbers et al. (2012).

[38] Liebig et al. (2010), Rowntree et al. (2020).

farm through cover cropping and grazing.[39] The farms that I am familiar with in Wisconsin that do rotational grazing and cover cropping don't seem to have nearly the level of success building soil carbon that Brown has (though they almost always have more organic matter than comparable soils in row crops).[40]

One key management difference became evident to me when I had the chance to visit Brown's farm in 2019. We walked to a section of the pasture that had been grazed the day before. What struck me was the amount of grass left in the pasture. In Wisconsin our guidance to grazing farmers used to be that they should move the animals by the time the grass was grazed down to 4 inches (10 cm) of height. The guidance had recently changed to telling farmers they should leave at least 15 cm and perhaps even 20 cm of grass in the paddock at the end of a grazing period. Wisconsin grazing educators were telling farmers that leaving that much residue behind in the paddock would allow the grass to regrow faster, so they would not really be reducing the total amount of feed for their animals, because they would be able to regraze that paddock sooner.[41]

The grass in Brown's paddock was flattened in many places, but the average height was closer to 30 cm than 20. This is the height when most Wisconsin farmers would think grazing should start, not when it should stop. I asked Brown what the recovery period would be for that paddock, and he replied that he would wait at least a year before putting cows back in that paddock. Given Wisconsin land prices, farmers cannot make money if they only graze their pastures once in a year. Depending on the soil and the weather, paddocks here are grazed three and sometimes four or even five times a year.

There is a similar gap between what Brown does with cover crops and what happens on the farms I know. The Wisconsin farms that are really committed to cover crops have winter wheat in their cropping rotation. The market for winter wheat is not as well supported as the maize or soybean market in Wisconsin, but winter wheat is harvested in July, while soybeans and maize are harvested in September or October. Harvesting the wheat in July gives cover crops time to grow while the weather is still warm. The only cover crops that can grow after mid-September in Wisconsin are rye or triticale, and unless we have an unusually warm fall even they don't grow much that late in the year.

Farmers in Wisconsin and elsewhere are experimenting with ways to plant cover crops while the main cash crop is still growing, but the competition between crops still slows the growth of the main crop or the cover crop or both, especially if there is not enough rain. In addition to planting the usual fall covers, the Brown ranch also puts a full year devoted to cover crops into their cropping rotation. They get some economic benefit from the cover crop if they graze it, and Brown says the soil health benefits compensate for the lost sales by reducing his costs for fertilizer and weed and pest management. However, for most farmers devoting a quarter or a fifth of

[39] Montgomery (2017), Brown (2018).

[40] Sanford et al. (2012).

[41] Paine and Jackson (2020).

their best cropland to cover crops seems out of reach economically, especially where land prices are high.

7.7 Variation in Soils and Climate

The amount of carbon that can be stored in soils also depends heavily on climate and soil type.[42] Sandy soils don't hold as much carbon as silt loams because the large particles have relatively little surface area for the carbon to bind to. And climate matters a lot. Soils in the humid tropics are notoriously low in carbon because the warm humid conditions support fast degradation of organic matter. In contrast, the relatively cool, dry nights and cold winters in a place like North Dakota slow or stop the activity of the soil organisms that consume the carbon the plants put in the soil. One of the frightening aspects of climate change is that as nights get warmer and the atmosphere holds more water vapor, the ability of soils to store carbon slows.

Where you have rich, deep soils, even as managed grazing adds carbon in the top 15 cm, carbon may be lost from deeper in the soil profile. That was the finding from a long-term trial comparing different cropping systems in southern Wisconsin. The maize (corn) – soybean cropping system lost the most carbon over the 20-year period, an average of 3.9 g of carbon per kg of soil, down to a depth of 1 meter. The rotational grazing system added 4 g of carbon per kg of soil in the top 15 cm of the soil profile over the 20-year period, but it lost carbon between 15 and 100 cm depth, for a net carbon loss of 0.1 g per kg of soil.[43]

Clearly, the grazing system is still much better for soil carbon than the annual row crops, and it also has lower greenhouse gas emissions from fertilizer and pesticide applications and machinery use. However, this study indicates that we need to be cautious about assuming that near-surface increases in soil carbon provide a full picture of soil carbon fluxes.

7.8 Feeding the World

What does this mean for the potential of grazing to sequester soil carbon? Put simply, it means that to build significant amounts of soil carbon farmers will need to manage first for soil carbon and health, and only second for production. In turn that means that if we take grazing seriously as a tool to mitigate climate change we cannot expect it to produce as much meat and milk as it could if we manage grazing systems to maximize production.

---

[42] Gy and Jackson (2000).

[43] Sanford et al. (2012).

In the arguments over which system is better, industrial or grass-based agriculture, advocates of industrial livestock production often say that their system can produce more meat and milk per hectare, and so is needed to feed the growing world population. Advocates of grazing often respond by pointing out that with the right management grazing can produce similar amounts of meat and milk on the same land type. When it tries to match the production of industrial agriculture, however, grazing often reduces the amount of carbon stored in the soil and tends to increase nitrous oxide and other greenhouse gas emissions.[44]

But producing American amounts of meat for the whole world should not be our goal, and in fact, we shouldn't even be producing American amounts of meat for the U.S. As we discussed in Chaps. 2, 3, and 5, we cannot produce American amounts of meat for the whole world without serious climate impacts. And although American farmers already produce more meat and milk than Americans want to buy, the excess does not go to the hungry people of the world. Much of the excess does not even get to the food insecure people in our own country, because they can't afford to buy it.[45] Instead of trying to extend the current meat consumption patterns of the U.S. to the whole world our goal needs to be creating an agriculture that sustains natural resources while feeding people a healthy, culturally appropriate, but not excessive diet.

7.9 Closing Thoughts

A version of grazing that replicates the soil-building action of migratory wild ruminant populations, as Teague and Apfelbaum describe in Chap. 6, can deliver meat and milk without net greenhouse gas emissions. But there is a wide gap between the ecology of wild ruminants in North America before European settlement and most modern grazing systems. With current populations and land ownership patterns we cannot fully replicate the prehistoric ecosystems of wild ruminants, but we can come much closer than the standard contemporary grazing management system.

To reach that state, however, we need to make the economics of a carbon-friendly grazing system work for farmers, farmworkers, and eaters. Can we think about creative approaches for using livestock to produce food for humans, build soil carbon, restore biodiversity, and also treat people fairly? I think we can, but it will take

---

[44] Maximizing meat and milk production means more animals and therefore higher enteric methane emissions per hectare. It also likely requires more fertilization and the associated greenhouse gas emissions to maintain the forage required to feed the cattle. Because of these emission sources, highly stocked grazing systems may be less effective greenhouse gas sinks than lightly or moderately stocked grasslands, even if soil carbon gains are greater in the more intensive grazing system (Liebig et al., 2010; Rowntree et al., 2020; Stanley et al., 2018).

[45] There are some efforts to use excess livestock production to feed the food-insecure within the U.S. Notable examples include the milk included in free and subsidized school lunches and during the past year the Coronavirus Food Assistance Program. However, excellent as those programs are, they fail to provide either real food security or reliably profitable prices for meat and dairy farmers.

sustained and focused effort and honest negotiation to shift our society's goals for agriculture. We will need to acknowledge and address the messiness and tradeoffs of how soils store and lose carbon and how livestock and grazing management affect greenhouse gas emissions to develop helpful and realistic grazing guidance for farmers, as well as policies that support regenerative grazing systems.

Bibliography

Adair, C. E., Barbieri, L., Shiavone, K., & Darby, H. M. (2018). Manure application decisions impact nitrous oxide and carbon dioxide emissions during non-growing season thaws. *Soil Science Society of America Journal, 83*, 163–172.

Aguirre-Villegas, H. A., Larson, R., & Reinemann, D. J. (2015). Effects of management and co-digestion on life cycle emissions and energy from anaerobic digestion: Modeling and analysis: Effects of management and co-digestion on life cycle emissions and energy from anaerobic digestion. *Greenhouse Gases: Science and Technology, 5*(5), 603–621. https://doi.org/10.1002/ghg.1506

Aguirre-Villegas, H. A., Passos-Fonseca, T. H., Reinemann, D. J., Armentano, L. E., Wattiaux, M. A., Cabrera, V. E., Norman, J. M., & Larson, R. (2015). Green cheese: Partial life cycle assessment of greenhouse gas emissions and energy intensity of integrated dairy production and bioenergy systems. *Journal of Dairy Science, 98*(3), 1571–1592. https://doi.org/10.3168/jds.2014-8850

Aguirre-Villegas, H. A., Passos-Fonseca, T. H., Reinemann, D. J., & Larson, R. (2017). Grazing intensity affects the environmental impact of dairy systems. *Journal of Dairy Science, 100*(8), 6804–6821. https://doi.org/10.3168/jds.2016-12325

Beauchemin, K. A., Ungerfeld, E. M., Eckard, R. J., & Wang, M. (2020). Review: Fifty years of research on rumen methanogenesis: Lessons learned and future challenges for mitigation. *Animal, 14*, s2–s16. https://doi.org/10.1017/S1751731119003100

Briske, D. D., Sayre, N. F., Huntsinger, L., Fernandez-Gimenez, M., Budd, B., & Derner, J. D. (2011). Origin, persistence, and resolution of the rotational grazing debate: Integrating human dimensions into rangeland research. *Rangeland Ecology & Management, 64*(4), 325–334. https://doi.org/10.2111/REM-D-10-00084.1

Brown, G. (2018). *Dirt to soil one family's journey into regenerative agriculture*. Chelsea Green Publishing.

Butterbach-Bahl, K., Baggs, E. M., Dannenmann, M., Kiese, R., & Zechmeister-Boltenstern, S. (2013). Nitrous oxide emissions from soils: How well do we understand the processes and their controls? *Philosophical Transactions of the Royal Society B: Biological Sciences, 368*(1621), –20130122. https://doi.org/10.1098/rstb.2013.0122

Cain, M., Lynch, J., Allen, M. R., Fuglestvedt, J. S., Frame, D. J., & Macey, A. H. (2019). Improved calculation of warming-equivalent emissions for short-lived climate pollutants. *Npj Climate and Atmospheric Science, 2*(1), 29. https://doi.org/10.1038/s41612-019-0086-4

Capper, J. L. (2012). Is the grass always greener? Comparing the environmental impact of conventional, natural and grass-fed beef production systems. *Animals, 2*(4), 127–143. https://doi.org/10.3390/ani2020127

Capper, J. L., Cady, R. A., & Bauman, D. E. (2009). The environmental impact of dairy production: 1944 compared with 2007. *Journal of Animal Science, 87*(6), 2160–2167. https://doi.org/10.2527/jas.2009-1781

Dangal, S. R. S., Tian, H., Xu, R., Chang, J., Canadell, J. G., Ciais, P., Pan, S., Yang, J., & Zhang, B. (2019). Global nitrous oxide emissions from pasturelands and rangelands: Magnitude, spatiotemporal patterns, and attribution. *Global Biogeochemical Cycles, 33*(2), 200–222. https://doi.org/10.1029/2018GB006091

Dini, Y., Gere, J., Briano, C., Manetti, M., Juliarena, P., Picasso, V., Gratton, R., & Astigarraga, L. (2012). Methane emission and milk production of dairy cows grazing pastures rich in legumes or rich in grasses in Uruguay. *Animals, 2*(2), 288–300. https://doi.org/10.3390/ani2020288

Fargione, J. E., Bassett, S., Boucher, T., Bridgham, S. D., Conant, R. T., Cook-Patton, S. C., Ellis, P. W., Falcucci, A., Fourqurean, J. W., Gopalakrishna, T., & Gu, H. (2018). Natural climate solutions for the United States. *Science Advances, 4*(11).

Feldt, T. (2020). Growing struggle over rising demand_ how land use change and complex farmer-grazier conflicts impact grazing management in the Western highlands of Cameroon. *Land Use Policy, 14*.

Franzluebbers, A. J., Paine, L. K., Winsten, J. R., Krome, M., Sanderson, M. A., Ogles, K., & Thompson, D. (2012). Well-managed grazing systems: A forgotten hero of conservation. *Journal of Soil and Water Conservation, 67*(4), 100A–104A. https://doi.org/10.2489/jswc.67.4.100A

Frith, S. (2020). The evidence for holistic planned grazing. In *Green meat?: Sustaining eaters animals and the planet* (pp. 98–106). McGill-Queen's University Press.

Garnett, T., Godde, C., Muller, A., Röös, E., Smith, P., & de Boer, I. (2017). Grazed and Confused? *Ruminating on cattle, grazing systems, methane, nitrous oxide, the soil carbon sequestration question – and what it all means for greenhouse gas emissions*, p. 127. https://www.leap.ox.ac.uk/article/grazed-and-confused-ruminating-cattle-grazing-systems-methane-nitrous-oxide-soil-carbon-sequ

General Mills. (n.d.). *Study: White Oak pastures beef reduces atmospheric carbon*. http://www.csrwire.com/press_releases/41932-Study-White-Oak-Pastures-Beef-Reduces-Atmospheric-Carbon

Gerber, P. J., Hristov, A. N., Henderson, B., Makkar, H., Oh, J., Lee, C., Meinen, R., Montes, F., Ott, T., Firkins, J., Rotz, A., Dell, C., Adesogan, A. T., Yang, W. Z., Tricarico, J. M., Kebreab, E., Waghorn, G., Dijkstra, J., & Oosting, S. (2013). Technical options for the mitigation of direct methane and nitrous oxide emissions from livestock: A review. *Animal, 7*(s2), 220–234. https://doi.org/10.1017/S1751731113000876

Global Warming Potential Values. (n.d.). https://www.ghgprotocol.org/sites/default/files/ghgp/Global-Warming-Potential-Values%20%28Feb%2016%202016%29_1.pdf

Gy, E. G. J., & Jackson, R. B. (2000). The vertical distribution of soil organic carbon and its relation to climate and vegetation. *Ecological Applications, 10*(2), 14.

Herrero, M., Henderson, B., Havlík, P., Thornton, P. K., Conant, R. T., Smith, P., Wirsenius, S., Hristov, A. N., Gerber, P., Gill, M., Butterbach-Bahl, K., Valin, H., Garnett, T., & Stehfest, E. (2016). Greenhouse gas mitigation potentials in the livestock sector. *Nature Climate Change*. https://doi.org/10.1038/nclimate2925

Hristov, A. N., Oh, J., Firkins, J. L., Dijkstra, J., Kebreab, E., Waghorn, G., Makkar, H. P. S., Adesogan, A. T., Yang, W., Lee, C., Gerber, P. J., Henderson, B., & Tricarico, J. M. (2013). SPECIAL TOPICS—Mitigation of methane and nitrous oxide emissions from animal operations: I. A review of enteric methane mitigation options1. *Journal of Animal Science, 91*(11), 5045–5069. https://doi.org/10.2527/jas.2013-6583

IPCC. (2014). *Climate change 2014: Synthesis report. Contribution of working groups I, II and III to the fifth assessment report of the Intergovernmental Panel on Climate Change* (p. 151). https://www.ipcc.ch/report/ar5/syr/

IPCC. (2021). *Climate change 2021: The physical science basis. Contribution of working group I to the sixth assessment report of the Intergovernmental Panel on Climate Change*. Cambridge University Press. https://www.ipcc.ch/report/ar6/wg1/?emci=3b983f0e-3cfc-eb11-b563-501ac57b8fa7&emdi=9cd148db-58fc-eb11-b563-501ac57b8fa7&ceid=1994345#SPM

Jackson, R. D., Oates, L. G., Schacht, W. H., Klopfenstein, T. J., Undersander, D. J., Greenquist, M. A., Bell, M. M., & Gratton, C. (2015). Nitrous oxide emissions from cool-season pastures under managed grazing. *Nutrient Cycling in Agroecosystems, 101*(3), 365–376. https://doi.org/10.1007/s10705-015-9685-6

Jakobsen, M., Preda, T., Kongsted, A., & Hermansen, J. (2015). Increased foraging in outdoor organic pig production—Modeling environmental consequences. *Food, 4*(4), 622–644. https://doi.org/10.3390/foods4040622

Katz-Rosene, R. M. (2020). How do livestock impact the climate. In *Green meat?: Sustaining eaters animals and the planet* (pp. 43–63). McGill-Queen's University Press.

Kriegl, T., & McNair, R. (2005). *Pastures of plenty. Center for integrated agricultural systems*. University of Wisconsin-Madison. https://www.cias.wisc.edu/wp-content/uploads/2008/07/pastplenty607.pdf

Li, C., Frolking, S., & Butterbach-Bahl, K. (2005). Carbon sequestration in arable soils is likely to increase nitrous oxide emissions, offsetting reductions in climate radiative forcing. *Climatic Change, 72*(3), 321–338. https://doi.org/10.1007/s10584-005-6791-5

Liebig, M. A., Gross, J. R., Kronberg, S. L., & Phillips, R. L. (2010). Grazing management contributions to net global warming potential: A long-term evaluation in the northern Great Plains. *Journal of Environment Quality, 39*(3), 799. https://doi.org/10.2134/jeq2009.0272

Lyon, A., Bell, M. M., Gratton, C., & Jackson, R. (2011). Farming without a recipe: Wisconsin graziers and new directions for agricultural science. *Journal of Rural Studies, 27*(4), 384–393. https://doi.org/10.1016/j.jrurstud.2011.04.002

Maasakkers, J. D., Jacob, D. J., Sulprizio, M. P., Turner, A. J., Weitz, M., Wirth, T., Hight, C., DeFigueiredo, M., Desai, M., Schmeltz, R., Hockstad, L., Bloom, A. A., Bowman, K. W., Jeong, S., & Fischer, M. L. (2016). Gridded National Inventory of U.S. methane emissions. *Environmental Science & Technology, 50*(23), 13123–13133. https://doi.org/10.1021/acs.est.6b02878

MacLeod, M., Gerber, P., Mottet, A., Tempio, G., Falcucci, A., Opio, C., Vellinga, T., Henderson, B., & Steinfeld, H. (2013). *Greenhouse gas emissions from pig and chicken supply chains: A global life cycle assessment*. Food and Agriculture Organization of the United Nations. http://www.fao.org/3/i3460e/i3460e00.htm

Mitloehner, F. (2020, September 3). For methane, GWP100 not measuring up. *Clarity and leadership for environmental awareness and research at UC Davis*. https://clear.sf.ucdavis.edu/blog/methane-gwp100-not-measuring

Montes, F., Meinen, R., Dell, C., Rotz, A., Hristov, A. N., Oh, J., Waghorn, G., Gerber, P. J., Henderson, B., Makkar, H. P. S., & Dijkstra, J. (2013). SPECIAL TOPICS—Mitigation of methane and nitrous oxide emissions from animal operations: II. A review of manure management mitigation options1. *Journal of Animal Science, 91*(11), 5070–5094. https://doi.org/10.2527/jas.2013-6584

Montgomery, D. (2017). *Growing a revolution*. W.W. Norton & Co.

Nichols, K. L., Del Grosso, S. J., Derner, J. D., Follett, R. F., Archibeque, S. L., Stewart, C. E., & Paustian, K. H. (2016). Nitrous oxide and methane fluxes from cattle excrement on C3 pasture and C4-dominated shortgrass steppe. *Agriculture, Ecosystems & Environment, 225*, 104–115. https://doi.org/10.1016/j.agee.2016.03.026

Oates, L. G., Undersander, D. J., Gratton, C., Bell, M. M., & Jackson, R. D. (2011). Management-intensive rotational grazing enhances forage production and quality of subhumid cool-season pastures. *Crop Science, 51*(2), 892. https://doi.org/10.2135/cropsci2010.04.0216

Oteros-Rozas, E., Ontillera-Sánchez, R., Sanosa, P., Gómez-Baggethun, E., Reyes-García, V., & González, J. A. (2013). Traditional ecological knowledge among transhumant pastoralists in Mediterranean Spain. *Ecology and Society, 18*(3), art33. https://doi.org/10.5751/ES-05597-180333

Paine, L., & Jackson, R. (2020). *Understanding soil carbon dynamics in pasture systems*. Green Lands Blue Waters. https://greenlandsbluewaters.org/wp-content/uploads/2020/10/GLBW-Soil-carbon-pastures-fact-sheet-LPaine-RJackson-Sep2020-FINAL.pdf

Rowntree, J. E., Ryals, R., DeLonge, M. S., Teague, W. R., Chiavegato, M. B., Byck, P., Wang, T., & Xu, S. (2016). Potential mitigation of Midwest grass-finished beef production emissions with soil carbon sequestration in the United States of America. *Future of Food: Journal on Food, Agriculture and Society, 4*(3), 31–38. https://www.thefutureoffoodjournal.com/index.php/FOFJ/article/view/78/71

Rowntree, J. E., Stanley, P. L., Maciel, I. C. F., Thorbecke, M., Rosenzweig, S. T., Hancock, D. W., Guzman, A., & Raven, M. R. (2020). Ecosystem impacts and productive capacity of a multi-species pastured livestock system. *Frontiers in Sustainable Food Systems, 4*, 13. https://www.frontiersin.org/articles/10.3389/fsufs.2020.544984/full

Sanford, G. R., Posner, J. L., Jackson, R. D., Kucharik, C. J., Hedtcke, J. L., & Lin, T.-L. (2012). Soil carbon lost from Mollisols of the North Central U.S.A. with 20 years of agricultural best management practices. *Agriculture, Ecosystems & Environment, 162*, 68–76. https://doi.org/10.1016/j.agee.2012.08.011

Savory, A. (2013). How to fight desertification and reverse climate change. https://www.ted.com/talks/allan_savory_how_to_fight_desertification_and_reverse_climate_change/transcript?language=en#t-1217846

Schlesinger, W. H., & Amundson, R. (2019). Managing for soil carbon sequestration: Let's get realistic. *Global Change Biology, 25*(2), 386–389. https://doi.org/10.1111/gcb.14478

Spratt, E., Jordan, J., Winsten, J., Huff, P., van Schaik, C., Jewett, J. G., Filbert, M., Luhman, J., Meier, E., & Paine, L. (2021). Accelerating regenerative grazing to tackle farm, environmental, and societal challenges in the upper Midwest. *Journal of Soil and Water Conservation, 76*(1), 15A–23A. https://doi.org/10.2489/jswc.2021.1209A

Stanley, P. L., Rowntree, J. E., Beede, D. K., DeLonge, M. S., & Hamm, M. W. (2018). Impacts of soil carbon sequestration on life cycle greenhouse gas emissions in Midwestern USA beef finishing systems. *Agricultural Systems, 162*, 249–258. https://doi.org/10.1016/j.agsy.2018.02.003

Starrs, P. F. (2018). *Transhumance as antidote for modern sedentary stock raising* (p. 12).

Teague, R., Provenza, F., Kreuter, U., Steffens, T., & Barnes, M. (2013). Multi-paddock grazing on rangelands: Why the perceptual dichotomy between research results and rancher experience? *Journal of Environmental Management, 128*, 699–717. https://doi.org/10.1016/j.jenvman.2013.05.064

U.S. Department of Agriculture. (2016a). *U. S. Agriculture and Forestry Greenhouse Gas Inventory, 1990–2013* (Technical Bulletin No. 1943). US Department of Agriculture. https://doi.org/10.15482/USDA.ADC/1264344

U.S. Department of Agriculture. (2016b). *U. S. Agriculture and Forestry Greenhouse Gas Inventory, 1990–2013*. U.S. Department of Agriculture, Office of the Chief Economist, Global Change Program Office. https://www.usda.gov/sites/default/files/documents/USDA_GHG_Inventory_1990-2013_9_19_16_reduced.pdf

U.S. Department of Agriculture Economic Research Service. (2020). *Meat supply and disappearance: Meat production, imports, total supply, exports, total and per capita disappearance with history.* https://www.ers.usda.gov/data-products/livestock-meat-domestic-data/livestock-meat-domestic-data/#All%20supply%20and%20disappearance

U.S. Department of Agriculture Economic Research Service. (2021). *Livestock & meat domestic-data.* https://www.ers.usda.gov/data-products/livestock-meat-domestic-data/

U.S. Environmental Protection Agency. (2020). *Inventory of U.S. Greenhouse Gas Emissions and Sinks 1990–2018.* https://www.epa.gov/ghgemissions/sources-greenhouse-gas-emissions

U.S. Environmental Protection Agency. (n.d.-a). Clean water act Section 404 and agriculture. *Section 404 of the Clean Water Act.* Retrieved September 3, 2021, from https://www.epa.gov/cwa-404/clean-water-act-section-404-and-agriculture

U.S. Environmental Protection Agency. (n.d.-b). *Livestock anaerobic digester database.* Retrieved March 11, 2021, from https://www.epa.gov/agstar/livestock-anaerobic-digester-database

U.S. Environmental Protection Agency. (n.d.-c). Nonpoint source: Agriculture. *Polluted Runoff: Nonpoint Source (NPS) Pollution.* Retrieved September 3, 2021, from https://www.epa.gov/nps/nonpoint-source-agriculture

USDA-NASS. (2019a). *2017 census of agriculture, table 8. Farms, land in farms, value of land and buildings, and land use: 2017 and 2012.* US Department of Agriculture. https://www.nass.usda.gov/Publications/AgCensus/2017/Full_Report/Volume_1,_Chapter_2_US_State_Level/

USDA-NASS. (2019b). 2017 census of agriculture, Table 43. *Selected Practices: 2017 and 2012*. US Department of Agriculture. https://www.nass.usda.gov/Publications/AgCensus/2017/Full_Report/Volume_1,_Chapter_2_US_State_Level/

VanderZaag, A. C., Jayasundara, S., & Wagner-Riddle, C. (2011). Strategies to mitigate nitrous oxide emissions from land applied manure. *Animal Feed Science and Technology, 166–167*, 464–479. https://doi.org/10.1016/j.anifeedsci.2011.04.034

Wattiaux, M. A., Uddin, M. E., Letelier, P., Jackson, R. D., & Larson, R. A. (2019). Invited review: Emission and mitigation of greenhouse gases from dairy farms: The cow, the manure, and the field. *Applied Animal Science, 35*(2), 238–254. https://doi.org/10.15232/aas.2018-01803

Winders, B., & Ransom, E. (2019). Introduction to the global meat industry: Expanding production, consumption, and trade. In *Global meat: Social and environmental consequences of the expanding meat industry*. MIT Press.

Wolf, B., Zheng, X., Brüggemann, N., Chen, W., Dannenmann, M., Han, X., Sutton, M. A., Wu, H., Yao, Z., & Butterbach-Bahl, K. (2010). Grazing-induced reduction of natural nitrous oxide release from continental steppe. *Nature, 464*(7290), 881–884. https://doi.org/10.1038/nature08931

Zubieta, Á. S., Savian, J. V., de Souza Filho, W., Wallau, M. O., Gómez, A. M., Bindelle, J., Bonnet, O. J. F., & de Faccio Carvalho, P. C. (2021). Does grazing management provide opportunities to mitigate methane emissions by ruminants in pastoral ecosystems? *Science of the Total Environment, 754*, 142029. https://doi.org/10.1016/j.scitotenv.2020.142029

Chapter 8
Lightening Our Carbon Hoofprint

Diane Mayerfeld

Abstract The relationship between livestock agriculture and climate is complex, with major differences between livestock types and production approaches; regional variation in climate, soils, alternative land use, and culture; and a good deal of uncertainty. As a result, solutions that do not take this variation into account are unlikely to have the promised effect. However, the fact that there is no one simple solution to the climate impacts of livestock has a strong silver lining, which is that everyone can take action to reduce their food and agriculture greenhouse gas emissions – their carbon hoofprint – including anyone who eats and anyone who farms. This chapter focuses on actions individual eaters and farmers can take, from vegans to meat lovers, and from concentrated animal feeding operation farmers to committed graziers.

Keywords Carbon footprint · Carbon hoofprint · Climate change · Climate-smart agriculture · Diet · Food waste · Greenhouse gas emissions · Livestock · Meat consumption · Soil carbon · Soil health

8.1 Introduction

What have we learned from the preceding chapters? If nothing else, they show that the relationship between livestock agriculture and climate is complex, with major differences between livestock types and production approaches; regional variation in climate, soils, alternative land use, and culture; and a good deal of uncertainty. As a result, simplistic solutions are unlikely to have the promised effect, even if they were politically feasible. Complexity is not the same as chaos, though. Here are three key themes that have emerged:

D. Mayerfeld (✉)
Division of Extension, Agriculture Institute, University of Wisconsin, Madison, WI, USA
e-mail: dbmayerfeld@wisc.edu

D. Mayerfeld (ed.), *Our Carbon Hoofprint*, Food and Health,
https://doi.org/10.1007/978-3-031-09023-3_8

1. **Livestock generate a lot of greenhouse gas emissions world-wide, but livestock also play a critical role in sustainable agricultural systems**

Chapter 2 explains how compared to staple grains and pulses, the production of meat and other livestock products emits more greenhouse gases per unit of protein or calories. This of course is the finding that drives the whole debate around livestock and climate change. In general, beef and lamb have the highest greenhouse gas emissions, while poultry has the lowest emission intensity per unit of meat.

As Chap. 3 lays out, however, at the same time livestock can play a critical role in making agriculture overall more sustainable and in making marginal lands produce food. Ruminants like cows and sheep are particularly effective at using crops that build soil health and fertility, and can also thrive on land that is too steep or rocky or cold or dry to grow staple plant crops that humans can eat. Thus, despite their high average emissions, we need cattle, sheep, and other ruminants to support a sustainable agriculture now and into the future. The question is, how do we raise them in a way that minimizes climate change and other environmental impacts and that supports people?

2. **On average, modern industrial livestock production has lower greenhouse gas emissions per kilo of meat or milk produced than traditional grazing systems, but these calculations do not take soil carbon fluxes into account. Industrial, efficient livestock systems can also lead to overproduction and waste.**

There is a wide range in greenhouse gas emissions from livestock, and reducing the emission intensity (the amount of greenhouse gas emissions per unit of meat or milk) of animal products is a critical step toward reducing the global climate impacts of livestock. As Chap. 4 discusses, intensive North American and European livestock production systems tend to have lower emissions intensities, especially for beef and dairy, than extensive production systems in Asia, Africa, and South America. In large part these lower emissions intensities are a result of breeding animals for quick growth and high milk production, and of feeding them carefully balanced, high quality rations. Both of these factors allow animals to gain weight more quickly, with a lower total intake of feed. Because the animals don't live as long before they reach slaughter weight, they have fewer direct emissions in the form of manure and in the form of methane as a by-product of digestion. Because they consume less feed over their lifetime, it is often assumed that the indirect emissions from feed production are also lower. For these reasons one solution that has been proposed is to shift more livestock production from extensive grass-based or backyard systems to modern industrial facilities.

As Chap. 5 lays out, however, there are some important caveats to the potential for efficient industrial livestock systems to reduce the climate impacts of livestock. One is that the analyses that find that industrial systems have a lower emissions intensity for cattle do not consider the effects of soil carbon storage and emissions in their model. Another is that efficiency is associated with low cost, which leads to over-production, over-consumption, and waste, and so does not result in a net reduction of emissions.

3. **Grazing systems can store enough soil carbon to offset all the greenhouse gas emissions from livestock in those systems. However, most grazing livestock are managed to maximize meat production rather than carbon sequestration, and therefore have net greenhouse gas emissions. Managing for carbon storage and soil health would significantly reduce total meat production.**

Climate activists have noted that storing as much carbon as possible in the soil is an important pathway to help mitigate climate change.[1] As Chap. 6 discusses, grazing systems with diverse perennial vegetation can store much more carbon in the soil than the annual crops used for feed in industrial livestock production. Grasslands grazed by herds of wild ruminants built deep, carbon-rich soils over thousands of years, even as the landscape was managed so hunter-gatherer-farmers could harvest the ruminants for food and other products. Converting those grazing landscapes to cropland in North America and other parts of the world released much of that stored soil carbon to the atmosphere, and so contributed to the build-up of atmospheric CO_2 worldwide. Converting cropland that grows animal feed back to grazing land can potentially store enough carbon in the soil to completely offset or even exceed the manure and digestion greenhouse gas emissions from livestock on that land.

As Chap. 7 lays out, however, grazed grasslands don't always store carbon, and we still have a lot to learn about how to manage grazing systems to minimize net greenhouse gas emissions. In most cases management that minimizes net greenhouse gas emissions produces less meat and milk per unit of land.[2]

But just saying "it's complicated" doesn't move us forward much, so this chapter will try to lay out some positive steps forward. The fact that there is no one simple solution to the climate impacts of livestock has a strong silver lining, which is that *everyone* can take action to reduce their food and agriculture greenhouse gas emissions – their carbon hoofprint – including anyone who eats and anyone who farms. The rest of this chapter will focus on actions individual eaters and farmers can take, and the next chapter will look at possible policy approaches.

8.1.1 What Can Individual Eaters Do?

I will start by trying to answer my neighbor's question about how to choose a climate-friendly individual diet.

First and foremost, try to minimize *all* your food waste. Not only are the greenhouse gas emissions from growing, processing, storing, and transporting that food

[1] Hawken (2017), Bossio (2020), Bossio et al. (2020), Toensmeier (2016), Montgomery (2017).
[2] Rowntree et al. (2020).

going to waste, but if the discarded food goes to a landfill it will generate methane as it decomposes. Fortunately, recognition that food waste is an environmental problem is growing, so it is easy to find good advice on how to reduce food waste, from tips on how to avoid over-buying to recipes for using peels and bones to make stock.[3] It is also socially acceptable to bring home leftovers from a restaurant meal. In January 2020 when I pulled out my reusable food storage container to bring home half of the large chop served at the banquet for a grazing conference, I was curious and a little concerned about how my table-mates would view this behavior. To my pleasant surprise, rather than disapproving, several of the people at my table commented that they wished they had thought to bring a similar container.

Second, if you are comfortable reducing the amount of meat and dairy you eat, that will help reduce your overall carbon footprint, and every reduction helps, even if it is small. If a vegan diet works for you that is great. There are plenty of other people who will eat the animal products resulting from livestock production that supports sustainable agriculture.

Third, you can choose what type of meat to eat based at least in part on its carbon hoofprint. If you don't have an easy way to find out how the meat was produced, then poultry is the option with the lowest average greenhouse gas emissions per gram of protein. Pork is intermediate, and beef and lamb have the highest average greenhouse gas intensities. However, if you have access to sustainably raised pastured beef, that might be your most climate-friendly meat choice, especially if the meat is from a dairy steer or cow.[4] Keep in mind, though, that a grass-fed label doesn't guarantee that the farm is storing soil carbon. If you buy from local farms you may be able to talk to the farmer about their specific practices, and you can know that the meat did not contribute directly to tropical deforestation.

A fourth and perhaps most important thing the individual consumer can do for the climate is to speak out. If you buy meat directly from a farmer, let them know that you care about greenhouse gas emissions and soil health, and ask them if they are using the practices described below. If you are buying from a grocery or a restaurant tell the meat manager or server that you would like to know how the meat was raised, including its greenhouse gas emissions. And tell your local, state, and national elected representatives that you care about the climate impacts of agriculture, including livestock, and that you expect them to enact policies that protect the climate and help consumers make climate-friendly choices.

Finally, while doing what you can to minimize the carbon hoofprint of your diet is helpful, remember that for most North Americans and Europeans agriculture and diet are a relatively small component of our overall climate footprint. Changes in

[3] A quick web search reveals waste reduction tips from government agencies such as the U.S. Food and Drug Administration https://www.fda.gov/food/consumers/tips-reduce-food-waste and the U.S. Environmental Protection Agency https://www.epa.gov/recycle/reducing-wasted-food-home, as well as from nonprofit environmental and health organizations and media outlets.

[4] Meat from dairy herds has lower emissions than meat from beef herds because the cows are producing milk at the same time that they are bearing calves, so their manure and enteric and feed emissions are divided between the milk and meat they produce.

transportation, home energy improvements, and reductions in consumption of non-food goods can have a much greater impact on an individual's carbon footprint.

8.1.2 *What Can Farmers Do?*

"It's not the cow – it's the how." I first heard this catchy phrase in a speech by Nicolette Hahn Niman, and it has since caught my eye a number of times. It means that beef is not inherently either good or bad, but that its environmental impacts, including its climate impact, depend on how it is raised. However, the how is more complicated than "grazing is good and confinement is bad" (or vice versa). Both grazing and feedlot or confinement production can cause serious environmental damage and result in significant greenhouse gas emissions if they are poorly managed. Conversely, good management can minimize negative environmental impacts in both systems. Until now livestock management has focused primarily on cost-effective production and secondarily on reducing water pollution and local air pollution. Moreover, "the how" includes the question "how many?" because the scale of livestock production definitely affects livestock's impact on the climate, as well as on many other environmental goods, from clean water and air to the thousands of other species that sustain and delight us.

What steps can industrial livestock farmers take to reduce their emissions? There are three main areas that farmers with industrial or mixed livestock systems can manage to reduce their carbon hoofprint: how they handle manure, what they feed their animals, and how they grow the feed for their livestock. There is active research on the climate benefits of all three of these approaches, but the researchers and extension advisors I work with have hesitated to make management recommendations to farmers based on reducing greenhouse gas emissions unless those recommendations will at least be cost neutral for the farmer.

8.1.2.1 Manage Manure

All farms that raise animals indoors can focus on manure management. In the U.S. manure management before spreading accounts for about 13% of overall agricultural greenhouse gas emissions, and roughly 80% of those manure emissions come from dairy and swine manure stored in liquid form in manure lagoons.[5] One promising technology to reduce greenhouse gas emissions from manure is the use of biodigesters, which capture the methane emitted from anaerobic decomposition of the manure. Methane is the same thing as natural gas, so once it is contained and

[5] U.S. Environmental Protection Agency (2020). Although beef cattle numbers are higher than dairy cow populations, their manure has much lower methane emissions because it is stored in dry packs rather than in liquid form. Liquid manure storage facilities provide an ideal environment for methanogenic microbes.

stored it can be used for electricity generation, heating, or even vehicle fuel. Thus, manure digesters can reduce overall greenhouse gas emissions in two ways. First, the methane that is captured can replace fossil fuels that would otherwise be used to make electricity, fuel, or heat, and so reduce the emissions from producing and burning those fossil fuels. Second, when the methane is burned to generate electricity or heat the farm it changes into CO_2 and water vapor, and as we know, the warming power of CO_2 is much lower than that of methane. As we discussed in the last chapter, CO_2 lasts longer than methane, but much of this carbon dioxide is part of the biogenic carbon cycle. In other words, the carbon that is released to the atmosphere by burning methane generated from manure is mostly carbon that was taken out of the atmosphere within the last year or two by the plants the animal ate.[6] In contrast, CO_2 released by burning fossil fuels is carbon that has been stored in the Earth's crust for millions of years and would stay there indefinitely if it is not extracted to make electricity, heat, transportation fuel, fertilizer, or the myriad other services and products humans get from coal, oil, and natural gas.

The drawback of this manure digestion technology is that it is expensive to build and operate.[7] As a result, U.S. farms have been slow to install digesters. In Wisconsin more than 300 farms are designated as Concentrated Animal Feeding Operations, meaning they each have more than 700 cows (a few have more than 5000 cows). However, as of early 2022 the state only had 39 manure digesters, plus 6 under construction. So fewer than a sixth of Wisconsin's largest dairy farms use manure digesters. More than 2000 additional Wisconsin dairy farms have more than 100 cows each. Nation-wide the U.S. had just over 300 manure digesters early in 2022.[8] In contrast, Germany has 3000 manure digesters, which convert nearly half the livestock manure produced in that country into electricity and/or heat.[9]

Another manure management option is to cover the stored manure and burn off (flare) the methane that the manure generates. Although this approach does not

[6] The proportion of livestock greenhouse gas emissions that are biogenic depends on the production practices of the farm. In most cases farms with stored manure rely on fossil fuels to produce their fertilizer and power the equipment to plant and harvest feed, and so their greenhouse gas emissions include both biogenic and non-biogenic carbon. Many grass-based farms use some fossil fuel, but typically less than industrial or mixed livestock systems.

[7] While very small-scale manure digesters can be built relatively cheaply and have been adopted in smallholdings (Montes et al., 2013), the cost of building and managing a system that can handle the manure from even 50 cows escalates rapidly (Vance Haugen 2012, personal communication).

[8] Two of the Wisconsin digesters are community digesters, each serving three farms near the CAFO threshold, so the total number of CAFOs served may be slightly higher than the number of digesters. Twenty-two of the digesters constructed in Wisconsin received U.S. Department of Agriculture grants, and many digesters also received tax credits and other grant funding. (U.S. Environmental Protection Agency, 2015, n.d.-a, n.d.-b, n.d.-c)

[9] Oehmichen and Thrän (2017). Europe is the only region in the world where methane emissions declined in a global survey of methane sources, and it seems likely that Germany's commitment to manure digesters was a factor in that decline. The Global Methane Budget 2000–2017 (2020), https://www.globalcarbonproject.org/methanebudget/index.htm

replace fossil fuels used for electricity or heat, it can be very effective at reducing methane emissions from manure, and it is less expensive than a digester.[10]

Adjusting the timing of manure application may also slightly reduce total greenhouse gas emissions. Methane generation is higher when temperatures are warm, so land applying stored manure in spring or early summer can reduce emissions from stored manure during the warm summer months. Applying manure when plants are growing rapidly and taking up nitrates may reduce the risk of large nitrous oxide emissions. The benefits of different ways of applying manure to agricultural land are mixed. For example, incorporating the manure can reduce water quality problems from phosphorus runoff but may increase the risk of nitrous oxide emissions. There are other manure management options that may reduce emissions, such as aerobic composting and liquid – solid separation, though more research is needed on the full life-cycle impact of different manure management approaches.[11]

8.1.3 Adjust What the Animals Eat

Enteric emissions (methane produced in the course of ruminant digestion) are the biggest source of direct livestock greenhouse gas emissions both in the U.S. and worldwide. How much methane a cow or sheep produces is influenced both by her genetics and her diet. One advantage of industrial livestock systems is that they give farmers control over the feed their animals consume. Normally two factors dominate how farmers decide what rations to provide: the cost of the feed and how well it translates to animal production. If farmers also prioritize feeds that reduce methane emissions they could further reduce greenhouse gas emissions from industrial beef and dairy farms. The optimal rations will vary depending on what feeds are locally available and on the livestock type (lactating dairy cows, beef cattle, sheep, etc.). For example, several studies have estimated that by adjusting rations for dairy cows to include slightly more grain and less forage, farms in the Midwestern and Northeastern U.S. could reduce their enteric methane emissions by 15% or more.[12]

There is also considerable research on using small quantities of additives to different feed mixes to reduce methane emissions. For example, one recent study indicates that adding a particular type of dried kelp to cattle feed can reduce enteric emissions by more than 60%, at least in the short term. Other feed additives may also reduce enteric methane emissions, again in the short term, including

[10] In addition to cost, another advantage of manure covering and methane flaring is that it poses less risk of actually increasing total manure methane emissions. In contrast, manure digesters are designed to favor methane production, so it is critical that they be well-managed to prevent methane leaks and to use the methane they generate. (Veltman et al., 2018)

[11] Montes et al. (2013), Hristov et al. (2013), Aguirre-Villegas et al. (2015, 2017), Kim et al. (2019).

[12] However, increasing the proportion of grain even further could again raise methane emissions (Veltman et al., 2018; Wattiaux et al., 2019).

bromochloromethane, chloroform, and cyclodextrin.[13] These latter compounds raise some questions about animal welfare and human health impacts, while the use of kelp raises questions about the potential impact on marine environments, which can have implications for atmospheric carbon, as well as biodiversity and human food security.[14] So while feed additives are a promising area of investigation, more work is needed on the long-term value and other environmental, social, and climate impacts of specific compounds.

Another step farmers can take to reduce greenhouse gas emissions is to make sure their animals do not eat more protein than they need for healthy growth. Excess protein in the diet increases the amount of nitrogen in the animal's urine and dung, which can in turn lead to higher nitrous oxide emissions, both while the manure is being stored and after it is applied to crop fields.[15]

8.1.4 Reduce Emissions from Feed Production

As farm advisors look at optimizing feed mixes to reduce greenhouse gas emissions, they need to also keep in mind the greenhouse gas emissions produced in growing the feed. There are a number of steps farmers can take to reduce the emissions associated with feed production, starting with fertilizer management.

Nitrous oxide emissions from key feed crops such as maize and soybeans account for at least 10% of agricultural greenhouse gas emissions in the U.S. Most of these emissions are the result of nitrogen fertilizer applications to the crops. Manure applications can also result in nitrous oxide emissions. At the very least, farmers can avoid applying more nitrogen fertilizer than the crops need.

The challenge is that figuring out the right amount of fertilizer to apply is not easy. If weather conditions are favorable, then the crop can use more nitrogen than if the weather is too dry or hot or cold. Nitrogen also easily dissolves in water and may leach away or run off the field before the plant can use it. The point where adding more fertilizer will no longer yield a greater profit varies depending on the price of the fertilizer and the price for the crop, as well as the weather. Farm fields don't have uniform fertility, so unless the farmer has expensive "precision" machinery that can adjust fertilizer application rates based on information stored in its Global Positioning System, they will probably wind up applying too much fertilizer in some places and not as much as the crop could use in others.[16] For all these reasons farmers are often tempted to add more fertilizer than they should, because they don't want to risk losing yield if the weather is good and/or the price for the crop

---

[13] Roque et al. (2021), Kebreab and Roque (2021), Hristov et al. (2013).

[14] Sala et al. (2021).

[15] Hristov et al. (2013), Montes et al. (2013).

[16] Even with precision farming equipment, human error or equipment malfunction can result in too much fertilizer in some places and too little in others.

turns out to be high. At least in some cases climate change (in the form of increased severe rain in late spring) is one of the factors that causes farmers to apply excess nitrogen, as these quotes from Midwestern farmers illustrate.

> 'If it keeps raining and it's warm, we're going to lose nitrogen, big time lose nitrogen, and that's when you've got to come back in and put some more [nitrogen] on or you're going to lose the crop, and there's "why did you lose the crop?" when with another 10 to 15 gallon of [liquid nitrogen fertilizer] you can fix it'.

> 'We usually put [a little extra nitrogen on] just to make sure if we have a really wet year, like we had last year and how this year is turning out, that we still have some nitrogen left over [to ensure sufficient yields]'.[17]

Unfortunately, that approach results in excess nitrogen polluting water bodies, as well as rising levels of nitrous oxide in the atmosphere. In many cases, that approach also costs farmers money over time. So a clear course of action for farmers seeking to reduce their greenhouse gas emissions is to avoid the temptation to apply more than the recommended amount of fertilizer and to ensure that their machinery is adjusted to prevent overapplication.

In recent years there has been a welcome move to understanding that soil is a complex ecosystem and not merely a physical matrix that can hold plant roots, nutrients, and water. Farmers and researchers increasingly seek to improve the health and functioning of the soil ecosystem rather than simply optimizing the amount of phosphorus, potassium, and nitrogen it contains. It turns out that soil organic matter (carbon) is highly correlated with soil health, and the practices that improve soil health also reduce emissions of CO_2 from soils and even build soil carbon, at least near the surface. Following these soil health practices is a critical step farmers can take to reduce their greenhouse gas emissions. In brief these soil health principles are:

1. Reduce disturbance. The main source of soil disturbance is tillage (plowing), so reducing tillage as much as possible can help protect soil carbon as well as reducing the fossil fuel burned to run tillage equipment. Reducing tillage can also reduce nitrous oxide emissions. However, pesticides also disturb the soil ecosystem, and fully eliminating tillage often requires increased use of herbicides to manage weeds, so continuous no-till is not always the most sustainable choice.
2. Maximize soil cover. Keeping the soil covered prevents erosion and protects the soil – and the life in the soil – from extreme swings in moisture and temperature. It is important to maximize soil cover both during the growing season for the main crop and during the fallow season (winter in North America and Europe). Farmers can use mulch, including crop residue, and/or cover crops to maintain soil cover after the cash crop is harvested and early in the growing season before the main crop has grown enough to fully cover the ground.
3. Maximize living roots in the soil. There are several components to this principle. As with maximizing soil cover, one is temporal. Annual crops like maize and soybeans only grow for five or six months, and farmers need to plant cover crops to keep living roots in the ground during the months when the annual crops are not present. Another component is spatial. Some plants have shallow root systems while others send their roots deep,

[17] Houser et al. (2019).

and some develop more dense root mats than others. Planting a diverse mix of cover crops can help with expanding the spatial reach of living roots, at least for a short time.

4. Maximize biodiversity. Efficient modern cropping systems rely on monocultures. In the U.S. maize and soybeans occupy 57% of the harvested cropland, and in some regions such as the states of Iowa and Illinois more than 90% of cropland is in these two crops. To date, most research on intercropping or use of companion crops (growing two or more species close together in the same field) has had lower yields and higher costs compared to growing one crop at a time. However, farmers can introduce some biodiversity by including more crop types in their rotation, as well as by using cover crops. Extending the crop rotation to include more than one or two species usually improves overall yields and can also greatly increase the opportunities for diverse cover crops. Another great way to increase biodiversity is by incorporating small areas of diverse native species into crop fields. Researchers and farmers in Iowa have experimented with putting strips of diverse native prairie into regular farm fields. They found that when they put just ten percent of the field into these diverse prairie strips they could reduce runoff and erosion by up to 90% compared to similar fields managed with soil health practices like no-till.[18]

There are numerous testimonials from farmers who have found that after 4 or 5 years of applying these soil health principles they are able to reduce nitrogen fertilizer below the recommended levels or stop using it altogether but still get good yields.[19] The challenge is that yield reductions during those transitional years create financial difficulties. In addition, farmers have to skillfully implement a host of complex practices and navigate some conflicting advice to build the soil health and microbial populations needed to compensate for reduced nitrogen fertilizer.

For example, to prevent yield reductions in crops planted into a good cover crop stand, agronomists may recommend increasing nitrogen applications.[20] While these increased fertilizer applications prevent short-term yield losses and so support successful adoption of cover-cropping, they also probably reduce the full soil health benefits from the cover crop and increase greenhouse gas emissions.

Another common challenge is that most farming systems are not set up to support diverse crops. The local seed dealer doesn't carry seed for unusual crops and does not know what varieties are best for the soils and climate in that area; the local knowledge support system from neighboring farmers to the local university can't give detailed advice on how to grow the crop; and if despite these barriers the farmer successfully raises the less common crop there is no local buyer or processor for it, so the farmer has to figure out where to sell it and how to get it there. The social stigma that farmers may experience when their fields look bad during transition, or simply because their different approach may be viewed as an implicit criticism of conventional agriculture can also be a barrier.[21]

[18] Schulte et al. (2017).

[19] Montgomery (2017), Massy (2018).

[20] The additional nitrogen is needed because the soil microorganisms that decompose the dead cover crop use nitrogen, and so reduce the amount of nitrogen available to the growing cash crop (Grint et al., 2020).

[21] Massy (2018).

8.1.5 Improve Grazing Management

What steps can grazing farmers take to reduce the carbon hoofprint of their farms? Unlike with industrial farms, the manure from their animals is not concentrated in one location, so biodigestion is not feasible, and in any case, methane emissions from manure in pastures are likely to be very low. But they can take action to reduce enteric emissions and build soil health (and soil carbon), just as industrial farmers can.

8.1.5.1 Adjust What the Animals Eat

Although ranchers do not have the same detailed control over what their animals eat as farmers in industrial livestock systems, how they manage their pasture and range affects the forage species and quality, and so can influence the enteric emissions from their livestock. In general, good quality forage high in digestibility and with adequate protein will produce less methane in the animal's digestive system and will also promote faster weight gain and more milk production. Farmers and ranchers can manage the quality and type of forage their animals eat both by planting species with good quality and by managing the timing and duration of grazing events.

However, as discussed in Chap. 7, farmers want to avoid having pastures with more protein than their animals need. The excess protein in the forage the animals consume goes to their manure and urine, and from there it can be converted to nitrous oxide, an extremely potent greenhouse gas. There is also evidence that some plant compounds, such as certain condensed tannins, can reduce enteric methane emissions. Farmers may not think of high tannins as good because tannins can inhibit the ability to digest protein. However, some traditional forage species such as birdsfoot trefoil (*Lotus corniculatus*) have higher tannin levels and can reduce methane emissions. Including some plants with tannins in the grazing system may be a promising strategy, but more research is needed in this area.[22]

Another way grazing farmers can reduce greenhouse gas emissions is by offering some supplemental grain or other high-energy feed such as certain kinds of food processing waste. It is natural to assume that if grazing is good, then relying purely on grazing must be best. But adding a modest amount of concentrate to a ruminant's diet can both reduce enteric emissions and increase the animal's growth and milk production.

[22] Hristov et al. (2013), Min et al. (2020).

8.1.5.2 Manage Grazing in a Way That Stores Soil Carbon

The great advantage that grazing systems have compared to livestock systems that rely on feeding animals harvested annual crops is that perennial grasslands are inherently ahead of annual crops in storing soil carbon. Let's look at how perennial pastures – the primary feed source for grazing livestock – perform on the four principles of soil health compared to annual cropping systems, and how they could do even better.

Soil Health Principle 1: Reduce disturbance. Unlike annual grain crops used for feed in industrial livestock systems,[23] traditional rangeland is not tilled or fertilized or treated with pesticides – the main sources of disturbance to agricultural soils. Perennial pastures in temperate climates with plentiful rain are managed more intensively, and grazing farmers here in Wisconsin occasionally use tillage to renovate a pasture, spot-spray troublesome weeds like Canada thistle or multiflora rose with herbicides, or apply nitrogen to boost pasture growth. But that tillage occurs perhaps once every 10 or 20 years, compared to annually or even multiple times per year for annual crops like maize and soybeans. Similarly, fertilizer and pesticide use on intensively managed pasture are much lower than for annual crops in the same area.

The primary sources of disturbance to pasture and range come directly from the grazing animals: the pressure from their hooves, the defoliation of the plants when the animals eat, and the deposition of manure. As Chap. 6 lays out, natural grasslands have co-evolved with grazing animals to thrive and take up carbon under these types of disturbance, but the level and duration of the disturbance are critical. While we need more research on how to optimize soil carbon storage in grazing systems, we know that moving livestock frequently, leaving ample residual forage at the end of a grazing session, and allowing long recovery periods before regrazing an area will improve soil health and carbon levels.

Soil Health Principle 2: Maximize soil cover. Unlike annual crop fields, pasture and range have year-round plant cover. Moreover, rather than being planted in rows with bare soil between the plant rows early and late in the growing season, the grasses and forbs in healthy pastures cover all the soil surface. However, if the pasture is grazed too heavily or if there is an area where the animals congregate too long some soil in grazing systems can be exposed. The same actions that reduce soil disturbance will also prevent soil exposure: moving livestock frequently, leaving

[23] Maize is the primary feed crop in the U.S. and is used increasingly in European livestock production. But almost all grains consumed by humans are used in great quantities for livestock feed, including wheat, barley, oats, millet, and soybeans as well as maize or corn. From 2015 to 2020 roughly 46% of maize, 35% of sorghum, 52% of barley, and 47% of oats, rye, millet, and mixed grains grown outside the U.S. were used as animal feed. During that same period 45% of corn, sorghum, barley, and oats grown in the U.S. were fed to livestock, and most of the rest was made into ethanol (biofuel) or exported (U.S. Department of Agriculture, Economic Research Service, 2021a, b).

ample residual forage in a paddock at the end of a grazing session, and allowing long recovery periods before regrazing an area.

Soil Health Principle 3: Maximize living roots in the soil. Because pastures are dominated by perennial grasses they automatically have living roots in the soil year-round, in contrast to annual crops. But not all roots are the same. Some species have deep root systems, while others have roots concentrated near the soil surface. Moreover, even within the same species how the plant is managed above-ground affects root growth. Plants use the carbohydrates they get from photosynthesis to grow both above and below-ground. The more leaf area they have, the more they can photosynthesize. As a result, leaving ample residual forage in a paddock at the end of a grazing session, and allowing long recovery periods before regrazing an area will also encourage root growth. In addition, encouraging a mix of species will likely increase overall root matter.

Soil Health Principle 4: Maximize biodiversity. Whereas annual crops are usually grown in monocultures, at least in mechanized farming systems, pastures generally have a mix of grasses and legumes. However, the diversity of pasture and range can vary widely, from pastures dominated by two or three species with similar growth habits, to highly diverse grazing systems with dozens of species. Farmers can deliberately manage for more diversity by planting a wider mix of grasses and legumes, including both deep- and shallow-rooted species, and warm-season (C-4) and cool-season (C-3) grasses. Farmers can also experiment with a variety of different grazing mixes that do well under different conditions.[24]

Livestock farmers can increase pasture and range diversity even more by including shrubs and trees in grazing systems. Silvopasture, the practice of including trees in grazing systems, increases both below- and above-ground carbon storage, and as a result Project Drawdown lists silvopasture as the 11th most effective strategy to reduce climate change in both the short and long-term.[25] In addition to storing carbon in the soil and in the trunks of the trees, adding trees to grazing systems can help farms adapt to some of the impacts from climate change. The shade provided by trees reduces heat stress for grazing livestock during the day. When there is enough rain, shade from trees typically reduces the growth of grass and legumes in the understory, but in hot, dry conditions shade can reduce water stress and help maintain forage growth. The total quantity of grass produced in most silvopastures over time is generally lower than the grass growth in comparable open pasture, but there is great value to having a source of feed when open pastures fail. Farmers can also use the foliage from many trees and shrubs as fodder for their livestock, either as a

---

[24] For example, in Wisconsin grazing farmers mostly plant cool-season grasses and legumes from Europe. These fast-growing species usually crowd out native warm-season grasses when they are mixed together, but a mix of native warm-season grasses planted in a separate pasture can provide good forage later in the year and increase soil health. (Gene Schriefer, personal communication)

[25] Baah-Acheamfour et al. (2014, 2016), Patel-Weynand et al. (2018), Hawken (2017).

regular component of their diet or as an emergency feed source.[26] For example at a field day one Wisconsin farmer recounted how in the drought of 2012 his grazing cows would come running when they heard the sound of his chainsaw, to eat the leaves from the newly felled tree. And a number of tree and shrub species contain higher levels of tannins that may help reduce enteric methane emissions.

Although it offers great promise as a carbon-friendly sustainable livestock practice, with silvopasture as with everything else the "how" – the management details – matter. The quick way to establish silvopasture is to take an existing woodland, remove enough trees to get sunlight on the ground, and plant grass. This establishment method is likely to result in a loss of soil carbon and certain to result in a loss of above-ground carbon. Farmers can have many good reasons to establish silvopasture this way, but reducing overall greenhouse gas emissions is not one of them.[27] Planting trees in open pasture, or planting trees and grass in a field that was in annual crops will almost certainly result in increased above- and below-ground carbon storage, although how much carbon is added will vary depending on the kind of trees and forages planted and the way they are managed.

8.1.5.3 Manage Nitrogen Use

Just as nitrogen fertilizer provides a quick boost to the growth of grains, it can also stimulate rapid, lush growth of perennial grasses. As a result, many grass farmers in North America and northern Europe sometimes apply synthetic nitrogen to their pastures. Reducing or eliminating applications of synthetic nitrogen to pastures can reduce the likelihood that microorganisms in pasture soils will convert some of that fertilizer nitrogen to the potent greenhouse gas nitrous oxide, and also reduces the greenhouse gas emissions from manufacturing synthetic nitrogen. Including some legumes in the pasture (but not too many) along with the manure deposited by the grazing animals should provide enough nitrogen for good pasture, as long as it is not overgrazed.

[26] Toensmeier (2016), Gabriel (2018). Feeding tree and shrub foliage is a common traditional practice in many parts of the world but was discouraged in North America and much of Europe for most of the past century or more with the rise of professional forest management and the intensification of livestock production. In recent years, however, rising interest in regenerative agricultural systems, including agroforestry, has prompted both farmer and researcher interest in the topic.

[27] However, from a carbon storage perspective, it is still better for farmers to establish silvopasture by thinning existing woodland than for them to cut all the trees down to create new open pasture. Mayerfeld et al. (2016), Blanco et al. (2019), Gabriel (2018), Broom et al. (2013).

8.1.6 The Best of Both Systems

The discussion of how to reduce greenhouse gas emissions from livestock has too often been framed as a competition between completely industrial production and a pure grazing approach. As we have seen, there is wide variation within each of those approaches in the greenhouse gas emissions they produce, and so there are many ways both industrial and grazing farmers can reduce greenhouse gas emissions within their existing systems. However, we also need to consider whether and when mixing elements from grazing and industrial production can offer further climate benefits.

In dairy a mixed approach that combines grazing with some grain supplementation appears to result in lower greenhouse gas emissions than either a complete indoor system or a system where the cows eat only grass. When researchers modeled greenhouse gas emissions from dairy farms in Wisconsin, they found that the scenario with the lowest emissions per unit of protein and fat corrected milk was the one where grazing cows were supplemented with grain to optimize their milk production.[28] They came to this conclusion even without including soil carbon in their model; adding the impacts of pasture on soil carbon would likely make the grazing dairy with grain supplementation look even better.

Many grazing dairies already provide grain supplements, though a few avoid all grain in order to market their products as purely grassfed. (Although purely grassfed products have a slightly different nutritional profile that may offer some health benefits, on average they are probably less climate-friendly than dairy from pasture-based farms that use some grain.) On the other hand, very few U.S. industrial dairies allow their cows graze. However, a few large intensive dairies are experimenting with letting their dry cows and heifers graze, and are finding that doing so reduces costs, improves the health of the cows, and complements soil health practices like adding cover crops and perennial pasture to the farm.[29]

Another example where mixing systems has the potential to reduce total carbon emissions is by integrating beef and dairy production. Unlike Norway, in the U.S. about 80% of beef comes from herds whose only purpose is to grow animals for meat.[30] Even in a state like Wisconsin, which prides itself on its many dairy farms and calls itself "the dairy state," most beef comes from specialized beef herds. The emissions intensity of beef from dairy calves is lower than that from pure beef herds, because a dairy cow's emissions help produce milk as well as calves. Rather than maintaining beef cow herds, grazing farms in places like Wisconsin could raise male dairy calves and surplus dairy heifers, with some grain supplements.

[28] Aguirre-Villegas et al. (2017).

[29] Adam Abel personal communication (2018).

[30] U.S. Department of Agriculture Economic Research Service (2021a, b), U.S. Department of Agriculture NASS (2021).

8.1.7 Advocate

These steps are not easy for farmers to take, but fortunately there is growing support for these actions at all levels, from local farm groups to national farm policy.[31] Farmers who make these changes can significantly reduce the carbon hoofprint of their own farms and may also inspire other farmers to make similar improvements. Finally, just as with individual eaters, one of the most significant actions individual farmers can take is to speak out about climate change and the positive actions the farm community can take.

Farmers can contact their political representatives, from the local Soil and Water Conservation Board, to the state legislature, and their congressional representative and senators in the U.S. They can support soil health policies in the local chapters of their conventional farm organizations and advocate for more support for climate-friendly conservation practices with the local Natural Resources Conservation Service Technical Committee. And they can share their thoughts on what agriculture can do to mitigate climate change or improve soil health with family and neighbors. The agricultural community in the U.S. has been politically conservative in recent decades, so talking about climate change may appear difficult and risky, but there are increasing signs of openness to at least some discussion about steps agriculture can take to mitigate climate change. Advocating for improved soil health is less controversial than referring to climate change and can be a good place to start.

And there are progressive farm organizations that are open to tackling climate change. Here Marbleseed in Wisconsin the Wisconsin Farmers Union, GrassWorks, the Savanna Institute, and are a few examples of farm groups that work to reduce agriculture's climate impact. Similar organizations exist in other states, and many of those groups welcome farmer members from across the country.[32]

Both farmers and consumers have many options to reduce their contributions to livestock's carbon hoofprint. If we move beyond defending or blaming livestock agriculture to looking at how each of us can best reduce our own greenhouse gas emissions, we can have a greater impact, both on our emissions, and in our credibility with others. Advocating for change is an important part of those efforts, especially if that advocacy is grounded in a sound understanding of the impact of livestock on greenhouse gas emissions and soil carbon, and if it engages honestly and respectfully with other groups.

---

[31] Noble Research Institute (n.d.), NRCS (2021a, b). One great model that has emerged in Wisconsin and some other states is for groups of farmers to support each other in doing this work. Five or more farmers in the same watershed can apply for a little bit of state funding to help them try actions that reduce water pollution and build soil health. Fortunately, almost all of these actions also reduce greenhouse gas emissions, though that was not the intent of the enabling legislation. This program has inspired farmers to try a range of sustainable practices, from reducing their use of tillage and planting cover crops to fine-tuning nitrogen applications (DATCP, n.d.).

[32] There are surely hundreds of environmentally and socially progressive farm groups just in the U.S., and apologies to the many not mentioned here (NSAC, n.d.).

Bibliography

Aguirre-Villegas, H. A., Larson, R., & Reinemann, D. J. (2015). Effects of management and co-digestion on life cycle emissions and energy from anaerobic digestion: Modeling and analysis: Effects of management and co-digestion on life cycle emissions and energy from anaerobic digestion. *Greenhouse Gases: Science and Technology, 5*(5), 603–621. https://doi.org/10.1002/ghg.1506

Aguirre-Villegas, H. A., Passos-Fonseca, T. H., Reinemann, D. J., & Larson, R. (2017). Grazing intensity affects the environmental impact of dairy systems. *Journal of Dairy Science, 100*(8), 6804–6821. https://doi.org/10.3168/jds.2016-12325

Baah-Acheamfour, M., Carlyle, C. N., Bork, E. W., & Chang, S. X. (2014). Trees increase soil carbon and its stability in three agroforestry systems in Central Alberta, Canada. *Forest Ecology and Management, 328*, 131–139. https://doi.org/10.1016/j.foreco.2014.05.031

Baah-Acheamfour, M., Carlyle, C. N., Lim, S.-S., Bork, E. W., & Chang, S. X. (2016). Forest and grassland cover types reduce net greenhouse gas emissions from agricultural soils. *Science of the Total Environment, 571*, 1115–1127. https://doi.org/10.1016/j.scitotenv.2016.07.106

Blanco, J., Sourdril, A., Deconchat, M., Ladet, S., & Andrieu, E. (2019). Social drivers of rural forest dynamics: A multi-scale approach combining ethnography, geomatic and mental model analysis. *Landscape and Urban Planning, 188*, 132–142. https://doi.org/10.1016/j.landurbplan.2018.02.005

Bossio, D. (2020, November 30). Solid ground: Earth's soils reveal climate, biodiversity & food security solutions. *Nature.Org*. https://www.nature.org/en-us/what-we-do/our-insights/perspectives/soils-revealed-climate-biodiversity-food-solutions/?src=s_two.exc.x.x.&sf133754315=1

Bossio, D. A., Cook-Patton, S. C., Ellis, P. W., Fargione, J., Sanderman, J., Smith, P., Wood, S., Zomer, R. J., von Unger, M., Emmer, I. M., & Griscom, B. W. (2020). The role of soil carbon in natural climate solutions. *Nature Sustainability, 3*(5), 391–398. https://doi.org/10.1038/s41893-020-0491-z

Broom, D. M., Galindo, F. A., & Murgueitio, E. (2013). Sustainable, efficient livestock production with high biodiversity and good welfare for animals. *Proceedings of the Royal Society B: Biological Sciences, 280*(1771), –20132025. https://doi.org/10.1098/rspb.2013.2025

DATCP. (n.d.). *Producer-led watershed Protection Grants impact report FY 2018–19*. Wisconsin Department of Agriculture, Trade and Consumer Protection. https://datcp.wi.gov/Pages/Programs_Services/ProducerLedImpactReport.aspx

Gabriel, S. (2018). *Silvopasture: A guide to managing grazing animals, forage crops, and trees in a temperate farm ecosystem*. Chelsea Green Publishing.

Grint, K., Smith, D., Arneson, N., DeWerff, R., Conley, S., & Werle, R. (2020, April 24). 2020 considerations for cover crop termination. *Wisconsin Weed Science*. https://www.wiscweeds.info/post/2020-considerations-for-cover-crop-termination/

Hawken, P. (Ed.). (2017). *Drawdown: The most comprehensive plan ever proposed to reverse global warming*. Penguin. https://www.drawdown.org/solutions-summary-by-rank

Houser, M., Gunderson, R., & Stuart, D. (2019). Farmers' perceptions of climate change in context: Toward a political economy of relevance. *Sociologia Ruralis, 59*(4), 789–809. https://doi.org/10.1111/soru.12268

Hristov, A. N., Oh, J., Firkins, J. L., Dijkstra, J., Kebreab, E., Waghorn, G., Makkar, H. P. S., Adesogan, A. T., Yang, W., Lee, C., Gerber, P. J., Henderson, B., & Tricarico, J. M. (2013). SPECIAL TOPICS Mitigation of methane and nitrous oxide emissions from animal operations: I. A review of enteric methane mitigation options. *Journal of Animal Science, 91*(11), 5045–5069. https://doi.org/10.2527/jas.2013-6583

Kebreab, E., & Roque, B. (2021, March 17). Feeding cows a few ounces of seaweed daily could sharply reduce their contribution to climate change. *The Conversation*. https://theconversation.com/feeding-cows-a-few-ounces-of-seaweed-daily-could-sharply-reduce-their-contribution-to-climate-change-157192

Kim, D., Stoddart, N., Rotz, C. A., Veltman, K., Chase, L., Cooper, J., Ingraham, P., Izaurralde, R. C., Jones, C. D., Gaillard, R., Aguirre-Villegas, H. A., Larson, R. A., Ruark, M., Salas, W., Jolliet, O., & Thoma, G. J. (2019). Analysis of beneficial management practices to mitigate environmental impacts in dairy production systems around the Great Lakes. *Agricultural Systems, 176*, 102660. https://doi.org/10.1016/j.agsy.2019.102660

Massy, C. (2018). *Call of the reed warbler: A new agriculture, a new earth*. Chelsea Green Publishing.

Mayerfeld, D., Rickenbach, M., & Rissman, A. (2016). Overcoming history: Attitudes of resource professionals and farmers toward silvopasture in Southwest Wisconsin. *Agroforestry Systems, 90*(5), 723–736. https://doi.org/10.1007/s10457-016-9954-7

Min, B. R., Solaiman, S., Waldrip, H. M., Parker, D., Todd, R. W., & Brauer, D. (2020). Dietary mitigation of enteric methane emissions from ruminants: A review of plant tannin mitigation options. *Animal Nutrition, 6*(3), 231–246. https://doi.org/10.1016/j.aninu.2020.05.002

Montes, F., Meinen, R., Dell, C., Rotz, A., Hristov, A. N., Oh, J., Waghorn, G., Gerber, P. J., Henderson, B., Makkar, H. P. S., & Dijkstra, J. (2013). SPECIAL TOPICS—Mitigation of methane and nitrous oxide emissions from animal operations: II. A review of manure management mitigation options1. *Journal of Animal Science, 91*(11), 5070–5094. https://doi.org/10.2527/jas.2013-6584

Montgomery, D. (2017). *Growing a revolution*. W.W. Norton & Co.

Noble Research Institute. (n.d.). *What is regenerative agriculture?* Noble Research Institute. Retrieved November 3, 2021, from https://www.noble.org/regenerative-agriculture/

NRCS. (2021a). *NRCS conservation programs: Conservation stewardship program (CSP)*. US Department of Agriculture, Natural Resource Conservation Service. https://www.nrcs.usda.gov/Internet/NRCS_RCA/reports/fb08_cp_cstp.html

NRCS. (2021b). *NRCS conservation programs: Environmental quality incentives program (EQIP)*. US Department of Agriculture, Natural Resource Conservation Service. https://www.nrcs.usda.gov/Internet/NRCS_RCA/reports/fb08_cp_eqip.html

NSAC. (n.d.). *National sustainable agriculture coalition members*. National Sustainable Agriculture Coalition. Retrieved November 3, 2021, from https://sustainableagriculture.net/about-us/members/

Oehmichen, K., & Thrän, D. (2017). Fostering renewable energy provision from manure in Germany – Where to implement GHG emission reduction incentives. *Energy Policy, 110*, 471–477. https://doi.org/10.1016/j.enpol.2017.08.014

Patel-Weynand, T., Bentrup, G., & Schoeneberger, M. M. (2018). Agroforestry: Enhancing resiliency in U.S. In *Agricultural landscapes under changing conditions: Executive summary* (WO-GTR-96a). U.S. Department of Agriculture, Forest Service. https://doi.org/10.2737/WO-GTR-96a

Roque, B. M., Venegas, M., Kinley, R. D., de Nys, R., Duarte, T. L., Yang, X., & Kebreab, E. (2021). Red seaweed (Asparagopsis taxiformis) supplementation reduces enteric methane by over 80 percent in beef steers. *PLoS One, 16*(3), e0247820. https://doi.org/10.1371/journal.pone.0247820

Rowntree, J. E., Stanley, P. L., Maciel, I. C. F., Thorbecke, M., Rosenzweig, S. T., Hancock, D. W., Guzman, A., & Raven, M. R. (2020). Ecosystem impacts and productive capacity of a multi-species pastured livestock system. *Frontiers in Sustainable Food Systems, 4*, 13. https://www.frontiersin.org/articles/10.3389/fsufs.2020.544984/full

Sala, E., Mayorga, J., Bradley, D., Cabral, R. B., Atwood, T. B., Auber, A., Cheung, W., Costello, C., Ferretti, F., Friedlander, A. M., Gaines, S. D., Garilao, C., Goodell, W., Halpern, B. S., Hinson, A., Kaschner, K., Kesner-Reyes, K., Leprieur, F., McGowan, J., … Lubchenco, J. (2021). Protecting the global ocean for biodiversity, food and climate. *Nature*. https://doi.org/10.1038/s41586-021-03371-z

Schulte, L. A., Niemi, J., Helmers, M., Liebman, M., Arbuckle, J. G., James, D. E., Kolkag, R. K., O'Nealh, M. E., Tomerf, M. D., Tyndalla, J. C., Asbjornseni, H., Drobneyj, P., Nealk, J., Van Ryswykl, G., & Wittec, C. (2017). Prairie strips improve biodiversity and the delivery of multiple ecosystem services from corn–soybean croplands. *Proceedings of the National Academy of Sciences, 114*(50), E10851. https://doi.org/10.1073/pnas.1719680114

Toensmeier, E. (2016). The carbon farming solution: A global toolkit of perennial crops and regenerative agriculture practices for climate change mitigation and food security. Chelsea Green Publishing.

U.S. Department of Agriculture, Economic Research Service. (2021a). *Feed grains yearbook tables*. Economic Research Service U.S. Department of Agriculture. https://www.ers.usda.gov/data-products/feed-grains-database/feed-grains-yearbook-tables/

U.S. Department of Agriculture, Economic Research Service. (2021b). *U.S. Bioenergy Statistics*. U.S. Bioenergy Statistics. https://www.ers.usda.gov/data-products/u-s-bioenergy-statistics/

U.S. Department of Agriculture NASS. (2021). *Livestock slaughter 2020 summary* (ISSN: 0499-0544; p. 68). National Agricultural Statistics Service.

U.S. Environmental Protection Agency. (2015). *Dane County community digester – Waunakee, WI*. https://www.epa.gov/sites/production/files/2016-05/documents/dane_county_agstar_site_profile_final_508_093015.pdf

U.S. Environmental Protection Agency. (2020). *Inventory of U.S. Greenhouse Gas Emissions and Sinks 1990–2018*. https://www.epa.gov/ghgemissions/sources-greenhouse-gas-emissions

U.S. Environmental Protection Agency. (n.d.-a). Clean water act Section 404 and agriculture. *Section 404 of the Clean Water Act*. Retrieved September 3, 2021, from https://www.epa.gov/cwa-404/clean-water-act-section-404-and-agriculture

U.S. Environmental Protection Agency. (n.d.-b). *Livestock anaerobic digester database*. Retrieved September 3, 2022, from https://www.epa.gov/agstar/livestock-anaerobic-digester-database

U.S. Environmental Protection Agency. (n.d.-c). Nonpoint source: Agriculture. *Polluted Runoff: Nonpoint Source (NPS) Pollution*. Retrieved September 3, 2021, from https://www.epa.gov/nps/nonpoint-source-agriculture

Veltman, K., Rotz, C. A., Chase, L., Cooper, J., Ingraham, P., Izaurralde, R. C., Jones, C. D., Gaillard, R., Larson, R. A., Ruark, M., Salas, W., Thoma, G., & Jolliet, O. (2018). A quantitative assessment of beneficial management practices to reduce carbon and reactive nitrogen footprints and phosphorus losses on dairy farms in the US Great Lakes region. *Agricultural Systems, 166*, 10–25. https://doi.org/10.1016/j.agsy.2018.07.005

Wattiaux, M. A., Uddin, M. E., Letelier, P., Jackson, R. D., & Larson, R. A. (2019). Invited review: Emission and mitigation of greenhouse gases from dairy farms: The cow, the manure, and the field. *Applied Animal Science, 35*(2), 238–254. https://doi.org/10.15232/aas.2018-01803

Chapter 9
Policy Pathways

Diane Mayerfeld

Abstract While individual actions are extremely important, they need to be supported by policy, both to achieve greater impact and to prevent unintended consequences. Because policy is so important, one of the most significant actions individuals can take is to advocate for good climate policy. But developing good policy is not simple. The first part of this chapter ruminates on ways to create policy that achieves our climate goals and avoids unintended negative consequences. Drawing in part on the research on commons management, this chapter proposes four guiding principles for policy development: (1) develop clearly articulated and shared goals, (2) base policy on good science, (3) distribute the benefits and burdens of the policy equitably and fairly, and (4) include ways to monitor and enforce compliance with the policy.

Keywords Climate change · Climate policy · Greenhouse gas emissions · Livestock

9.1 Introduction

The previous chapter lists actions individual farmers and eaters can take to reduce their carbon hoofprint, from changing manure management to reducing food waste. While individual actions are extremely important, they need to be supported by policy, both to achieve greater impact and to prevent unintended consequences. We saw an example of unintended consequences from individual actions in Chap. 3, where at the same time that per capita meat consumption in the U.S. declined (2005–2013), agricultural greenhouse gas emissions rose. The increase in greenhouse gas emissions came about because farmers and grain processors compensated for the decline in demand for feed grains for livestock by helping to pass the Energy

D. Mayerfeld (✉)
Division of Extension, Agriculture Institute, University of Wisconsin,
Madison, WI, USA
e-mail: dbmayerfeld@wisc.edu

D. Mayerfeld (ed.), *Our Carbon Hoofprint*, Food and Health,
https://doi.org/10.1007/978-3-031-09023-3_9

Policy Act of 2005, which required inclusion of ethanol in U.S. gasoline. That Renewable Fuel Standard resulted in soaring demand for maize to produce ethanol, which caused widespread land use change from conservation land to corn production and associated CO_2 and nitrous oxide emissions from soils.

Indeed, one of the values of individual actions is that they can prompt government to enact better policy. The small fringe of people who chose to ride their bikes to work and shop in the 1970s and 80s showed traffic planners in some cities that people might use bike paths, and in turn those bike paths have encouraged significant growth in bike commuting in cities like Portland, Oregon and Madison, Wisconsin. In the Netherlands, where bicycle infrastructure and other bike-friendly policies are really taken seriously, more than a quarter of all trips are by bicycle.[1] Similarly, those farmers who already manage their cropping and grazing systems for soil health are showing policy makers that certain farming practices can store carbon in soils, reduce greenhouse gas emissions and other negative environmental impacts, and still produce abundant healthy food.

Because policy is so important, one of the most significant actions individuals can take is to advocate for good climate policy. But developing good policy is not simple. The first part of this chapter ruminates on ways to create policy that achieves our climate goals and avoids unintended negative consequences. Based on that framework for policy development, I discuss a few of the many climate policies that have been proposed for livestock agriculture. Because the policy landscape is vast and changing rapidly, this will not be a comprehensive review. Finally, I suggest some policy ideas that have not been widely covered in the broad public conversation.

Drawing in part on the research on commons management I would like to propose four guiding principles for policy development: (1) develop clearly articulated and shared goals, (2) base policy on good science, (3) distribute the benefits and burdens of the policy equitably and fairly, and (4) include ways to monitor and enforce compliance with the policy.

9.1.1 Clear Policy Goals

First, in order to prevent unintended negative consequences good policy needs a clear and comprehensive vision. I propose at least four goals for policy makers to keep in mind as they seek to reduce the carbon hoofprint of livestock:

1. Agriculture should aim for net zero greenhouse gas emissions.
2. Agriculture needs to provide healthy, culturally appropriate food that is accessible to all.
3. Agriculture still needs to protect other environmental resources and ecosystem services, including biodiversity and clean water and air.

[1] Pucher and Buehler (2008), Buehler and Pucher (2012).

4. The costs and benefits of climate policy should be borne equitably by farmers, farm and food system workers, and eaters.

A few comments about each of these goals.

Why aim for net zero emissions? We have to aim at least for net zero because, as Chap. 5 discusses, even when emission intensity is low, total emissions can remain high when increasing consumption and waste counteract greater efficiency. Moreover, because greenhouse gas levels are already too high, reducing emissions a little bit is simply not enough. In that case, why not aim for net negative emissions? In the long run I agree that agriculture's goal should be to draw down atmospheric carbon at least by the amount agriculture has added through land use change and use of fossil fuel over the past two centuries. But the first step on this journey needs to be achievable as well as ambitious, and net zero seems like a good balance. I want to emphasize that this goal of net zero must be net zero within the agriculture of each region. Agriculture in general and the livestock sector in particular must avoid the disingenuous practice of using carbon sequestration in some other region or some other sector to offset its greenhouse gas emissions.[2]

As we focus on reaching zero emissions, agriculture still needs to feed people. And that food needs to be healthy. Grain production has much lower greenhouse gas emissions than production of fruits, vegetables, and livestock products, but a diet that is high in grains and deficient in produce, dairy, and meat already makes millions of people sick in the U.S. and around the world. Even in the wealthy U.S. millions of people cannot afford healthy food now; agricultural and food policy needs to make good food *more* accessible to all, not less. We also need to keep in mind that healthy food is connected to culture as well as quantities of vitamins, protein, and other nutrients.[3] Agriculture and food policy has to respect diverse food cultures at the same time that it seeks to discourage overconsumption and waste of livestock products and other food. Fortunately, focusing on food provides an opportunity to reduce agriculture's climate emissions and increase carbon storage. At present, nearly 40% of maize and 15% of soybeans grown in the U.S. are used to make ethanol and biodiesel for transportation fuel, and Brazil, Europe, and other regions likewise use staple crops to produce biofuels.[4] As transportation in the U.S. and around the world shifts to greater fuel efficiency and electric vehicles, we have an opportunity to devise policy that will encourage farmers to transition that land to uses that store carbon and do not emit greenhouse gases.[5]

[2] Foley (2021) lays out a cogent critique of easy and false pledges or claims of net zero emissions. Although he focuses on corporate greenwashing, other institutions, including non-profit universities, have engaged in similar questionable carbon accounting.

[3] See for example *In Defense of Food* by Michael Pollan and *The River is in Us* by Elizabeth Hoover.

[4] Bruckner et al. (2019), U.S. Department of Agriculture Economic Research Service (2021).

[5] Coming up with economically viable alternatives for this land currently growing grain for biofuel is critical. Some commodity groups are talking about increasing the required amount of ethanol in motor fuel in the U.S. as they anticipate an overall drop in demand for gasoline, although we know that the small climate gains from using biofuel rather than petroleum are outweighed by the effects of land use change (McFetridge 2021; Lark et al. 2015).

In many cases the same agricultural practices that reduce emissions or store carbon also benefit water and air quality and biodiversity. For example, reducing nitrogen applications will almost certainly reduce water pollution, as well as nitrous oxide emissions to the atmosphere; and converting cropland to perennial pasture or integrating strips of diverse perennials in cropland can improve water quality, store soil carbon, and increase biodiversity.[6] Nevertheless, it would be foolish to simply assume that all strategies to reduce agricultural greenhouse gas emissions will also improve other ecosystem services, so each agricultural climate policy should also be evaluated for its other environmental impacts.

Finally, policy needs to ensure that the burden of reducing agricultural emissions is shared fairly. Both worldwide and in the U.S. there are enormous inequities in the food system and in the costs of climate change. Some of those inequities briefly came into the spotlight during the pandemic of 2020, when food processing workers, especially workers of color in meat-processing plants, were disproportionately exposed to COVID-19.[7] Climate change also disproportionately affects farmworkers and less wealthy farmers who have to work outdoors in extreme weather, who are increasingly exposed to insects and disease vectors that are spreading as the climate warms, and who are less likely to have insurance or other programs to protect them from climate-related crop losses. Devising equitable policies to reduce the harm from climate change is both the right thing to do and also can expand political support for climate policy.

9.2 Good Science

Second, good policy has to be based on an accurate and shared understanding of the resource being managed.[8] That may sound obvious and non-controversial, but the continuing debate around meat and climate shows we are still far from a shared understanding of the role livestock plays in destabilizing or regulating climate. While most people agree that policy should be based on good science, no one can have a grasp of all the relevant research. Science can happen at many levels, and a lot of critical research happens at a reductionist scale. That is, in order to get a statistically significant result, the research attempts to hold all variables constant except the one or two factors being examined. Taken together this type of research can advance our understanding, but no individual study comes even close to explaining the full complexity of the relationship between agriculture and climate. In addition, often studies build on each other, creating clusters of research that have similar

[6] Kremen and Merenlender (2018).

[7] Hendrickson (2020), Saitone et al. (2021), Taylor et al. (2020).

[8] Wilson (2002), Ostrom et al. (1999). Although this requirement is absolutely foundational, it seems to get less attention in the literature on the commons than other factors such as the size and relationships of the community managing the commons and the ability to monitor resource use.

assumptions and boundaries, and not surprisingly come to similar conclusions, as we see in Chaps. 2, 4, and 6.[9]

This kind of insular research can promote polarization, as this quote from a conservative commentator indicates:

> Speaking of stupid, there's a study coming out of the University of Michigan which says that to meet the Biden Green New Deal targets, America has to, get this, America has to stop eating meat, stop eating poultry and fish, seafood, eggs, dairy, and animal-based fats. Ok, got that? No burger on July 4. No steaks on the barbecue. I'm sure Middle America is just going to love that.[10]

I am not aware of a study from the University of Michigan that says Americans have to stop eating meat, but as discussed in Chap. 3, numerous scholarly articles have suggested that eliminating livestock or at least beef production would be an effective strategy to combat climate change, based on simplistic assumptions.[11] In fact, neither Biden's climate proposals nor the Green New Deal legislation proposed by Representative Ocasio-Cortez even mention reducing meat production or consumption. Here is everything the Green New Deal bill says about agriculture:

> (1) it is the duty of the Federal Government to create a Green New Deal … [which] will require the following goals and projects …
>
> G) working collaboratively with farmers and ranchers in the United States to remove pollution and greenhouse gas emissions from the agricultural sector as much as is technologically feasible, including—.
>
> (i) by supporting family farming;
> (ii) by investing in sustainable farming and land use practices that increase soil health; and.
> (iii) by building a more sustainable food system that ensures universal access to healthy food[12];

This book attempts to lay a foundation for basing policy discussions on a better understanding of the resource by looking at both the limitations and the strengths of the different clusters of research cited by opponents of animal agriculture and by proponents of industrial livestock production and of holistic grazing. To reach an accurate and shared understanding of the problem we also need to view both biophysical and social processes both at the micro scale of a particular field, animal, or farm and at the macro scale of a region, nation, or even the whole world.

There are encouraging signs that the research around livestock and climate may be becoming slightly less polarized and selective. This recognition of nuance is an important first step toward a shared and accurate understanding of livestock's impact on the resource we need to manage in a global commons – our climate. Two recent

[9] This clustering of bodies of research also occurred at a small scale when I asked colleagues to suggest readings about the role of livestock in greenhouse gas emissions (Mayerfeld 2016).

[10] Fox Business (2021).

[11] Poore and Nemecek (2018), Garnett et al. (2017), Harwatt et al. (2017).

[12] Ocasio-Cortez (2019).

studies that reported significantly lower greenhouse gas emissions for grass-finished livestock on adaptive multi-paddock pasture than grain-finished meat also noted that it takes more than twice as much land to produce that grass-finished meat as meat from a feedlot. The Oxford Food Climate Research Network in the past promoted a message that livestock was the greatest source of anthropogenic greenhouse gases; in their 2020 report *Methane and the sustainability of ruminant livestock,* they acknowledge flaws in that analysis and instead emphasize that all sectors, including livestock, need to greatly reduce greenhouse gas emissions.[13] Yet as we see in the conservative quote above, the blunt argument against meat continues to influence the policy landscape by raising the specter that any form of supply management will set us on a slippery slope toward outlawing all livestock.

In addition, our scientific understanding of greenhouse gas emissions and carbon storage in agricultural systems is far from complete. Moreover, there is a great deal of variation in the effectiveness of practices to promote soil carbon storage or reduce greenhouse gas emissions between different sites. Given the urgency of the climate crisis we cannot let limited knowledge and variation stop us from moving forward with policy, but how can policy navigate between the shoals of uncertainty and urgency?

Fortunately, there are some livestock practices where the science is clear that they reduce net greenhouse gas emissions, and policy can begin by focusing on those practices. In addition, it is important to be open about scientific uncertainty and to be prepared to adjust policy as we learn more. Acknowledging uncertainty may seem a dangerous strategy when climate denialists have equated any uncertainty with reason to doubt all the science around climate change, but the danger of claiming certainty and then being proved wrong is greater than the risk that admitting some areas of uncertainty might prolong some climate skepticism. Scientists and climate activists can promote a 3-part message that

1) scientists are certain that burning fossil fuels, deforestation, and other human activities are changing the climate,
2) there are actions we can and must take now to reduce the severity of climate change, and
3) further research will help us continue to improve our understanding of specific dynamics of and solutions to climate change.

To deal with regional social and biophysical variation, good policy also has to take local context into consideration.[14] For example, a recent meta-analysis of management strategies to reduce emissions from beef production found that increasing feed efficiency was a promising strategy in some parts of the world, but not in the U.S. or Europe, where feed efficiency is already extremely high.[15] Moreover, even within countries context can differ both by geography and by social factors. The same

[13] Stanley et al. (2018), Rowntree et al. (2020), Lynch et al. (2020).

[14] See for example chapters 1, 8, and 9 in (Ostrom et al. 2002).

[15] Cusack et al. (2021).

study that found that improving feed efficiency offers little promise for reducing greenhouse gas emissions from beef in the U.S. identified improving management of grazed lands as the most promising strategy for reducing the carbon intensity of U.S. beef. But within the U.S. optimal management of grazing will look different in the dry Southwest or Northern Plains than in the humid Upper Midwest or Southeast, and the approaches that work for a small part-time cow-calf farm may be different than those for a large commercial herd.

9.3 Equity and Fairness

Third, good policy has to treat those affected equitably. In a traditional commons management negotiation around a pasture or a forest it might be possible to approach equity by including all the people who are directly affected in developing the rules for managing the resource, and much of the commons literature discusses the role of group size and cohesion in examples of successful traditional common resource management. With climate change the whole world is affected, and of course we cannot realistically negotiate policy with every human. Instead, it is critical to design a policy process that provides fair representation for all important stakeholder groups. To encourage equity, that representation needs to be based more on the number of people affected than on the economic power of different groups, and it also needs to include the voices of minority groups.

Power imbalances and the need for equity happen at many different scales: among nations, among businesses and sectors within a nation, among individuals. The policy discussion in this chapter will focus on the U.S., which is one of the world's largest greenhouse gas emitters, both by country and on a per capita basis. Both because of its high emissions, and as a wealthy country, the U.S. has an obligation to significantly reduce its emissions.[16] However, there are many different stakeholder groups within each country, and there are great differentials in the economic and political power of those groups. Ensuring that all farmers and all eaters are fairly represented in developing climate policy for livestock will not be easy, but it is vital that policy treat farmers and eaters of different wealth, different farming systems, and different cultures fairly.

[16] Union of Concerned Scientists (2020). From an equity perspective it makes more sense to look at per capita emissions than nation-wide emissions, but even that is complicated by factors like local climate and availability of renewable energy sources such as wind, solar, or hydro. Also, both national and per-capita emission figures can be distorted by exports and imports. For example, if the U.S. imports manufactured goods from China should the emissions from the production of those items count toward China's emissions or those of the U.S.?

9.4 Monitoring and Compliance

Fourth, good policy needs to be effective. The commons management literature stresses the importance of monitoring resource use and of the ability to impose sanctions on those who ignore the rules. This need for accountability brings us to a discussion of the role of regulation versus relying on voluntary action. Within the agriculture community in the U.S. the common wisdom is that "command and control" regulation is bad and that voluntary incentives are good. This belief is not new – the primary legislation to protect water quality, the Clean Water Act of 1972, exempted agricultural "nonpoint source" pollution from regulation.[17] However, the experience with the Clean Water Act suggests that relying entirely on voluntary incentives may not be effective. At the national, state, and local levels government has invested billions of dollars in voluntary conservation incentives for farmers designed to prevent soil erosion and water pollution. Despite these investments, erosion and nitrate and phosphorus leaching and runoff from agricultural fields continue to contaminate drinking water and make small and large water bodies unsafe for aquatic life and human recreation.[18]

Moreover, the distribution of voluntary incentives and support has often been deeply inequitable. In the U.S. the history of discrimination against Black and Native American farmers' access to federal loan programs is well documented.[19] Even beyond deliberate discrimination or implicit racial or ethnic bias by agency staff, the design of voluntary incentive programs usually favors the wealthy and well-connected over those farmers that the U.S. Department of Agriculture calls "socially disadvantaged." It is very difficult to apply for voluntary conservation assistance without a computer and good internet, as well as fluent English and good writing skills. It is also extremely helpful for applicants to visit in person with a representative of the agency managing the incentive program. Less wealthy farmers with full-time off-farm jobs may not be able to make such a visit during the work week when U.S. Department of Agriculture offices are open. They are also unlikely to have the time it takes to familiarize themselves with intricacies of the program and to fill out the complex application forms, on top of farming and working at and commuting to their off-farm job. Moreover, these incentive programs usually operate on a reimbursement basis, meaning they are only accessible to farms that have the money to spend on the improvement or practice up front.

Many voluntary programs also create unfairness when some farmers incur additional costs to protect the environment, while others continue to pollute. A common complaint from people in agriculture is that most farmers try to protect the environment, while a few bad actors cause most of the pollution and give all farmers a bad name. Currently, a major incentive program for conservation in agriculture, the

[17] U.S. Environmental Protection Agency (n.d.-a, n.d.-b, 404)

[18] Osmond et al. (2015), Gillon et al. (2016), Porter et al. (2015).

[19] Carpenter (2012), Rippon-Butler (2020), McCammon (2015), Newkirk (2019), Leonard et al. (2020), Mitchell (2005).

Environmental Quality Incentives Program (EQIP), prioritizes giving financial assistance for conservation to those farms that are causing a resource concern. In other words, if a county's EQIP funds are limited (as they always are) they will award those funds to the farmer whose management is causing soil erosion and fish kills rather than to the farmer whose land is well managed, and who is looking for some help to do even better management.[20] On one hand it makes sense to target limited funds to the worst problems, but on the other hand this approach may discourage some farmers from being proactive about protecting water and soil.

Elinor Ostrom, a preeminent scholar of common resource management, describes four ways of interacting with common resources. There are free-riders, people who think only of their own profit and do not cooperate to preserve common resources. Then there are people who are only willing to protect the resource if they are assured that they will not be exploited by free-riders. I suspect this second group represents most of us, both within and beyond the farming community. Third, there are people who limit their use of the resource in the hope that their actions will inspire others to do the same. And finally, there may be some genuine altruists who limit their use of common resources simply because it is the right thing to do.[21] Voluntary programs support the third and fourth groups who are already inclined toward protecting the common resource, but effective policy also has to make it more profitable for free-riders to comply with commons use restrictions than to flout the guidelines, and has to assure the second group (those willing to do their part if others are not taking advantage of them) that they will not be cheated by free riders.

Much of the research on the effectiveness of farm policy in achieving conservation goals focuses more on characteristics of landowners than on the detailed characteristics of the policy. When farmers are asked whether they prefer regulation or voluntary incentives most respond that voluntary incentives are preferable. However, it is unclear whether that response simply reflects prevailing assumptions that regulation is always burdensome. Contrary to the authors' expectations, a study of forest landowners in Sweden found strong acceptance of regulations requiring all landowners to follow certain environmental regulations, including a requirement to leave a minimum number of standing trees when harvesting timber. In the same survey landowners had more critical and mixed attitudes towards voluntary environmental incentive programs.[22]

---

[20] In 2019 fewer than half of the valid EQIP applications were funded (Congressional Research Service 2020). There is also a program that provides incentives to farmers who are good land conservationists, the Conservation Stewardship Program (CSP). Although the mandatory funding requirements for the CSP are only about half as high as those for EQIP, the actual program expenditures are now higher because they are cumulative. In other words, once a farmer enrolls in CSP they continue to be paid for their stewardship over time, whereas paying for an EQIP practice is a one-time reimbursement (NRCS 2021a, b).

[21] Ostrom et al. (1999).

[22] Danley (2019), Rissman et al. (2017), Upadhaya et al. (2021).

This is not to claim that regulations are always effective or equitable. Indeed, it is particularly important to ensure that regulation and enforcement are designed to be equitable and not impose excessive financial or management burdens. But the Swedish study shows that well-designed and fairly enforced regulation can be acceptable to those who are regulated, as well as effective in achieving conservation goals. As we seek to reduce livestock's carbon hoofprint we need to consider all the tools available, from "carrot" policies like education and voluntary incentives to "stick" policies such as taxes and regulations.[23]

9.5 Policy Ideas

Over the last year discussions of climate policy have burgeoned in the U.S. Despite a global pandemic dominating the news cycle and everyday life, the U.S. presidential campaign and the media covered climate change more in 2020 than in earlier years. Since the Biden administration took office in January 2021, media attention to climate and climate policy has remained high, and speculation about and suggestions for climate policy have become staples in the agricultural press.[24] This next section will look at some of the main policy ideas that have been proposed for the U.S.

9.5.1 *Voluntary Financial Incentives*

Most of the policy discussion focuses on voluntary incentives for farmers to increase soil carbon storage. One of many groups weighing in on potential climate policy for agriculture is the Food and Agriculture Climate Alliance (FACA), a coalition of farm and forest organizations, food businesses, and environmental advocates. Early in 2021 they released a document laying out a suggested framework for policy to reduce agriculture's climate impact. The document begins by listing three principles that guide their policy recommendations:

1. Support voluntary, market- and incentive-based policies.
2. Advance science-based outcomes.

[23] Monitoring and enforcement are also needed for voluntary incentive programs. The Renewable Fuels Standard specifies that maize grown on land converted from natural habitat can not be used for ethanol production subsidized by the act, but without monitoring and enforcement of that stipulation the act resulted in millions of acres of grassland and conservation land converted to maize production (Lark et al. 2015).

[24] See for example (Brooks 2021a; Cubbage 2021; Griffiths 2021).

3. Promote resilience and help rural economies better adapt to climate change.[25]

It is wonderful to see this coalition of groups come together to address climate change. (Some of the groups have a history of climate skepticism, and many have opposed each other in the past.) Their report contains many good recommendations, and their second principle, which advocates for science-based policy, aligns with this book's central premise that climate policy for agriculture has to be based on an accurate understanding of the impacts of livestock on greenhouse gas fluxes. But the unwritten corollary to their first principle greatly limits the ability of their recommendations to effect change. What is that unwritten corollary? It is "Oppose all (climate-related) regulation of private land management, especially in agriculture."

Their reliance on voluntary incentives risks exacerbating the existing inequities in American agriculture. Take for example FACA's lead recommendation to promote soil carbon sequestration, and think about how it might affect different farmers:

> Support a menu of voluntary federal options to encourage carbon sequestration, including: A performance-based tax credit for carbon sequestration modeled after 45Q [45Q provides a tax credit on a per-ton basis for qualified captured carbon dioxide].

The document goes on to recommend that

- The tax credit should be transferable, allowing maximum flexibility for participants.
- Relevant USDA agencies should play a significant consultative role in developing a policy guidance document covering measurement and verification that could be used for public incentives and by private markets.

Filing for a tax credit and providing the necessary measurement and verification of carbon sequestration are likely to be daunting barriers for many small farms. The farmers most likely to be able to take advantage of these incentives are large farms that have the financial resources to hire professional accounting and recordkeeping services, or farms owned by people with a background in finance or tax law. Creating a strong support structure to ensure that all farmers can navigate the program requirements would be extremely expensive.

The financial support offered to U.S. farmers during the COVID-19 pandemic offers an example of what is likely to happen under a voluntary financial incentive system without explicit and careful planning for equity. Under the initial round of the Coronavirus Food Assistance Program the top 5% of recipients received more than half of the money, while the bottom 10% of recipients received much less than

[25] Food and Agriculture Climate Alliance (2021). This group is not the only one focusing its climate policy recommendations for agriculture on voluntary incentives. From the U.S. Department of Agriculture to environmental organizations for now most of the policy discussion for agriculture centers on encouraging voluntary action. Mayer (2021), National Sustainable Agriculture Coalition (2019), U.S. Department of Agriculture (2021a).

1% of the funds (roughly 0.1%).[26] Those numbers actually understate the unequal receipt of aid, because they only show the distribution of assistance among those farm businesses that were able to successfully apply to the program; farmers who were not already used to working with the U.S. Department of Agriculture, had language barriers, or lacked internet access were largely left out altogether and do not appear in those statistics. The U.S. Department of Agriculture recently announced a supplementary aid program intended to reach some of the farmers who could not access the original assistance; it remains to be seen how effective the new efforts will be at reaching those farmers who have historically been excluded.[27]

One approach to consider for voluntary incentives is to target them to the practices, such as planted silvopasture and other perennial polycultures for agriculture, that we think are most effective at storing carbon and reducing emissions. This would be a change from current policy, which prioritizes practices such as no-till and cover crops, that farmers can adopt quickly and without changing the crops they grow, equipment they use, and markets they sell to; but which at best store modest amounts of carbon, unless the farm also adjusts the overall cropping system.[28]

9.6 Reducing Climate Impacts

The primary focus of this book and of much of the debate around livestock and climate is on reducing greenhouse gas emissions and so reducing the extent of climate change over the long term. However, agriculture also has the potential to influence some of the impacts of climate change. About half of all land in the U.S. and close to 40% of land worldwide is in agricultural use.[29] Because agriculture occupies so much of the world's land, especially near human populations, how that agricultural land is managed can ameliorate or exacerbate the effects of climate change. For example, one of the biggest effects of climate change in many areas, including my home state, has been more extreme rainfall events. Wisconsin has experienced a marked increase in extreme storms in recent years, and this trend is expected to persist.[30]

---

[26] Brasher et al. (2021).

[27] Shortly after the announcement of this special loan forgiveness program, a group of white farmers supported by conservative political groups filed suit claiming the program discriminates against them and so violates their constitutional rights (U.S. Department of Agriculture 2021b; Richmond 2021). On April 5th, 2021 the U.S. Department of Agriculture announced a grant program for organizations to help "socially disadvantaged" farmers and ranchers apply for pandemic assistance. Providing this kind of support is an important step in the right direction. Unfortunately, the deadline to apply was in 30 days, and the community groups we approached were unable to put together and submit a proposal in that time frame.

[28] Toensmeier (2016), Hawken (2017).

[29] Callahan (2019), Nickerson and Borchers (2012), World Bank (2018).

[30] Wisconsin Initiative on Climate Change Impacts (2022).

When intense rain falls on plowed land, much of it runs off and causes flooding, erosion and sedimentation, and destruction of infrastructure such as roads and bridges. In contrast, natural grasslands and forests can absorb much greater amounts of rain, so flooding, erosion, and property damage are less likely and less severe. Well-managed pasture can infiltrate nearly as much rain as natural grassland. Practices that increase soil organic matter, such as planting cover crops, can also improve the ability of soils to absorb rain. As a result, a recent climate task force in Wisconsin calls for converting grain crops to pasture and for increasing cover crops:

> To make Wisconsin's agricultural system more resilient to climate change, we need continuous living cover systems that allow farmers to decrease greenhouse gas emissions and increase the ability of soils to store carbon. The easiest solutions are to increase perennials (alfalfa hay, grass, and pastures) and cover crops (annual crops planted when fields are fallow) on agricultural land, include more rotational grazing as part of livestock operations so less grain is needed for animal feed, improve nutrient and manure management practices to reduce liquid manure and better align nutrient applications with plant nutritional needs, and avoid grassland or natural vegetation conversion to row crop production or urban development.
>
> We suggest Wisconsin promote practices to keep agricultural fields green all year with perennial vegetation and cover crops as a way to store more carbon in the soil, reduce soil erosion and nutrient runoff, and increase biodiversity. Marginal agricultural lands that are taken out of production could provide potential sites for installing solar infrastructure. These actions will increase resiliency to extreme weather events while improving environmental conditions, human health, and agriculture's overall resiliency to a changing climate. They will also support a move towards more sustainable energy production.[31]

As the recommendations above show, in many cases the same practices that reduce the impacts of climate change also reduce net greenhouse gas emissions. This positive synergy may not always be so clear, though. Take the example of silvopasture as a tool to reduce the risk from forest fires. The increasing severity of forest fires in the northwestern U.S. and elsewhere is in part a consequence of climate change. Converting forest near residential areas to silvopasture could lower the risk of severe fires by reducing the amount of fuel (shrubs, dead wood, dry grass) in the understory. However, if some trees are removed to encourage better grass growth, as is standard practice for converting forest to silvopasture, that would cause a loss of above-ground carbon storage. Still, that carbon loss would be much smaller than the loss a severe fire would cause.

9.6.1 Carbon Markets

No discussion of agricultural climate policy would be complete without considering carbon markets. This is the policy proposal that generates the most enthusiasm, the most skepticism, and the most confusion. The idea of a market that will pay farmers

[31] Wisconsin Initiative on Climate Change Impacts (2020).

for storing carbon is appealing to those who tend to distrust government. In a market that pays farmers to store carbon, farmers are the solvers of climate problems; not bad actors who are wrecking our climate and need to be forced to change their ways. In addition, farmers can choose whether or not to implement climate-friendly practices with a carbon market.[32]

But there is also considerable skepticism about relying on carbon markets. Some of this doubt stems from the experience with the Chicago Climate Exchange, which was established in 2003 and for a while offered farmers payments for switching to continuous no-till. However, late in 2010 the Exchange failed, leaving many farmers, climate activists, and climate skeptics disillusioned.[33] Other doubts are rooted in the difficulty of predicting and quantifying the amount of carbon stored by agricultural practices such as no-till and cover-cropping. Several studies have questioned the extent to which these practices store carbon over the long term, or if they just cause a temporary increase in organic matter near the surface that is later reversed or is exceeded by carbon losses deeper in the soil profile or increased nitrous oxide emissions.[34]

It is also worth noting that the notion that carbon markets are a way to address climate change without the heavy hand of government is misleading. A stable carbon market relies on government regulation in the form of a cap and trade program. The market may be facilitated by private companies, but it works by allowing an industry such as an electric utility or an airline to continue emitting greenhouse gases *above their regulatory cap* as long as they pay someone else to take an equal or greater amount of carbon out of the atmosphere. If these industries pay farmers for conservation practices that don't store as much carbon as predicted, then the market actually enables a continued build-up of greenhouse gases. In other words, if they are not based on reliable science and monitoring, carbon markets could prove to be a counterproductive climate policy tool.

The unanswered questions about how to measure and verify greenhouse gas reductions from agricultural conservation practices and about what a cap and trade program would look like have led to confusion. How will markets set prices? How will they verify how much carbon different practices store in different locations? How difficult will it be to access the markets? Will they only benefit large farms with degraded soils? What will the consequences be for landowners if it turns out their actions fail to store as much carbon as expected? Will they be required to refund the payments? How long will they be required to maintain a conservation practice? Right now different programs have different answers and are available in different areas. This diversity gives some farmers choice, but also increases the

---

[32] Bloomberg Editorial Board (2021), Brasher (2021), U.S. Department of Agriculture (2021a), Clayton (2021).

[33] Gronewold (2011), Morgan (2021).

[34] Li et al. (2005), Sanford et al. (2012), Dignac et al. (2017).

burden of research needed to decide whether to enter the market and which program to choose.[35]

Some of these questions can be addressed with thoughtful policy-making, but at its core a carbon market based on offsets does not move agriculture toward our goal of zero emissions. Under such a market any carbon stored by good agricultural practices is negated by enabling excess greenhouse gas emissions in other sectors. For livestock agriculture, participation in carbon markets would undermine the attempt to reach net zero emissions, because if the carbon stored in soils by good grazing or crop management is traded in a carbon market it cannot also be used to offset enteric and manure emissions and other agricultural greenhouse gas emissions.

9.6.2 Research

The knowledge gaps about the impacts of different farming practices on long-term storage of soil carbon and greenhouse gas emissions point to the need for much more research focused on agriculture and climate mitigation. At a minimum agricultural policy should dedicate a much larger proportion of research funding to mitigating climate impacts, and that research needs to focus both on how to reduce emissions and on how to store carbon over the long term. For example, grazing research has received much less funding and attention than the production of annual crops used for livestock feed, and most of the grazing research that has occurred has focused primarily on how grazing management can increase meat and milk production, rather than on minimizing enteric and nitrous oxide emissions and maximizing soil carbon storage.[36]

To address the goals of zero emissions and of feeding people we need research throughout the food system on the causes of food waste, on how much food is being wasted, and on strategies to minimize waste and better manage food that is discarded. For example, one of the traditional strengths of swine and poultry was their ability to eat food waste, from low quality or insect-damaged fruits and vegetables to processing wastes such as whey. Research on how to safely overcome barriers to feeding livestock food waste could help restore this function of livestock agriculture. At the same time, we need research on innovative and climate-friendly strategies to improve access to healthy food for all people.

We also need more long-term systems research that investigates both the biophysical and social aspects of real agricultural and food systems, rather than primarily rewarding reductionist experiments that may have statistical significance but limited applicability to the actual world of agriculture and food systems. One way

---

[35] Schlesinger and Amundson (2019), Cubbage (2021), Griffiths (2021), Castagné et al. (2020), Morgan (2021), Brooks (2021b).

[36] Academic research on grazing started with the work of Voisin in the middle of the twentieth century (See for example Voisin (1959)), but there was little grazing research in the U.S. until the 1990s.

to accomplish that is to include farmers and other food system stakeholders in research, from design to interpretation of findings, as the U.S. Department of Agriculture Sustainable Agriculture Research and Education Program encourages.[37] And if we are serious about tackling this issue, perhaps all government-funded agricultural research could be required to at least describe the climate implications of the project, in the same way that projects have to declare whether and how they include animals or human subjects, at least until our understanding of how to reduce net emissions increases substantially.

A number of voices have called for more research on developing meat substitutes that would be acceptable to eaters, from plant-based products like the "Impossible Burger" to lab-grown meat.[38] Research on these products needs to address their potential climate and health impacts, as well as tackling how to produce them economically. These products will likely supplement sustainably raised livestock products in meeting the world's demand for food, and we need to evaluate what their implications are for the concentration of agricultural business in the hands of large corporations, for human health, and for the environment.

9.6.3 Supply, Demand, and Regulation

Voluntary incentives, markets that pay farmers for soil carbon storage, better research – these three policy approaches for agriculture now have widespread attention and increasing support even in the agricultural sector. They are all worthwhile approaches to consider, but none of them addresses the underlying problem of overproduction of livestock and the resulting methane emissions. As we saw in Chap. 5, in the U.S. excess production of meat and dairy products leads not only to greater greenhouse gas emissions from farms and ranches, but also to low prices and recurring economic hardship for farmers, as well as greenhouse gas emissions from food waste.

While individual consumers, food businesses, and farms can take steps to reduce food waste, we also need to tackle this issue at a policy level. Since the Great Depression, U.S. farm and food policy has seesawed between encouraging maximum production to keep food accessible and preventing overproduction to protect farmer livelihoods. In recent decades the overall thrust of farm policy has been to encourage production and then attempt to protect farmer livelihoods by providing price supports through subsidized insurance, emergency payments to farmers, promoting export markets, and subsidized markets such as that for corn ethanol. This

[37] My position is partially funded by the U.S. Department of Agriculture SARE Program, and I often have to explain to researchers applying for a grant that when the Call for Proposals says "Research and Education projects include … significant farmer/rancher or other end-user involvement from the inception of the idea through the implementation of the project" that does not just mean that the farmer allows the researcher to run an experiment on their land (USDA n.d.).

[38] Klein (2021), Newman (2020), Tuomisto and Teixeira de Mattos (2011).

policy approach needs to be drastically re-examined with a frame of producing enough food to provide food security but not so much as to bankrupt farmers and encourage waste.

Here are a few suggestions for ways to shift food policy to a better balance between food security and overproduction, and to begin accounting for greenhouse gas emissions.

The first proposal is not new, and is likely to be popular with most American farmers. Country of Origin Labeling (COOL) requires grocery stores to provide information to buyers about where certain foods were produced. According to the U.S. Department of Agriculture, "products covered by the law include muscle cut and ground meats: lamb, goat, and chicken; wild and farm-raised fish and shellfish; fresh and frozen fruits and vegetables; peanuts, pecans, and macadamia nuts; and ginseng."[39] The original COOL legislation passed in 2002 and 2008 included beef and pork, but an appropriations bill passed in 2016 exempted beef and pork from these labeling requirements.[40] Restoring the requirement to show the country of origin for beef and pork and expanding the requirement to include retailers such as restaurants and food service would give consumers the ability to avoid buying beef imported from a place where it is likely to have contributed to tropical deforestation. This change would not affect the production of U.S. farms, but might reduce the incentive for excess production in other countries.

A second proposal is to end the checkoff programs that require all beef, pork, and dairy farmers to pay a fee to commodity organizations for marketing.[41] If farmers wish to engage in collaborative marketing they can do so, but government should not be in the business of marketing for private businesses, or of requiring farmers to participate in collective advertising programs.

In a related vein, the U.S. government can focus its dietary guidelines on human health rather than on marketing for specific agricultural sectors. The guidelines have made much progress on this front in the past 25 years, moving from implying that people only get protein from livestock products to now including plant-based protein sources. However, the continued listing of dairy as a critical food group seems rooted more in support of the dairy sector than in adult nutritional needs, even if the U.S. Department of Agriculture now lists fortified soy beverages as an option within that food group.[42]

Likewise, our government does not need to act as a marketing agent for our agricultural products on the international level. Our current policy on international trade in agricultural commodities is dominated by efforts to get other countries to buy our excess agricultural products, whether they need and want them or not, without

---

[39] U.S. Department of Agriculture (n.d.-a, n.d.-b)

[40] Agricultural Marketing Service (2016).

[41] Sanburg (2021), U.S. Department of Agriculture Agricultural Marketing Service (n.d.).

[42] U.S. Department of Agriculture and U.S. Department of Health and Human Services (2020), Food Forum et al. (2019).

consideration of climate implications.[43] International food trade can be vital for food security in the face of regional droughts, floods, or other disruptions, all of which are increasingly common with climate change, but it can also undermine rural economies and local food sovereignty. While export markets have occasionally benefitted American farmers, reliance on them has also caused severe financial damage, including the prolonged farm crisis of the 1980s. A trade policy that prioritizes food security and food sovereignty and also takes greenhouse gas emissions into account, rather than simply seeking to maximize exports, could benefit American farmers as well as the climate, especially if it is paired with domestic supply management policies.

Traditionally, there has been a lot of concern about supply management policies, though a few agricultural sectors, including potato and cranberry growers, have successfully collaborated to reduce overproduction. There are legitimate concerns that supply management could create barriers for new farmers or that it could raise food prices excessively. The link between overproduction of agricultural goods and climate change is not visible, while the suffering caused by insufficient food production is immediate, clear, and widespread. In contrast, the immediate economic costs of agricultural overproduction are either borne by a small fraction of the population, or they are not visible to the broader population because the cost of emergency payments to farmers is not differentiated from all the other taxes people pay. But given the urgency of addressing climate change, combined with the failure of government support of overproduction to either support thriving rural economies or healthy diets for all Americans, it is time to think about how to manage production to provide enough healthy food for everyone without growing far more than we need.

There are a number of possible policy tools for managing overproduction. One area to look at is reforming government-subsidized insurance programs that protect farmers against low prices. For example, in dairy, where overproduction has contributed to record numbers of farm bankruptcies in recent years, perhaps federal Price Loss Coverage should not be offered to new dairy farms with more than a certain number of cows, say 300 or 500 milking cows.

Another approach might tie subsidized insurance to carbon storage practices – a farm might only qualify for crop insurance (or other subsidies) if a minimum percentage of its land is in diverse perennial cover that builds soil carbon. The types of perennial cover that qualify would depend on the region, but they would have to meet the four principles of soil health: minimizing disturbance (pesticides and tillage), keeping the soil covered year-round, maximizing living roots in the soil, and having diverse plant groups. Examples of qualifying perennial polycultures could include strips of native prairie plantings or agroforestry practices such as diverse windbreaks or forested riparian buffers.[44] In contrast, current crop insurance

[43] For example, the U.S. has focused considerable effort on trying to force Canada to buy more U.S. dairy products, even though Canadian farmers have no trouble producing more than enough milk to meet Canadian demand (Laca 2020).

[44] Schulte et al. (2017), Patel-Weynand et al. (2018), Montagnini and Nair (2004).

programs issue payments based on past yield history of major commodity crops, and so indirectly encourage overuse of fertilizer and lack of cropping system diversity.

Increasingly, livestock in the U.S. are concentrated in very large animal feeding operations or CAFOs.[45] Requiring all new dairy and swine farms with more than 500 animal units (or another threshold) to reduce methane emissions from their stored manure through biodigestion could be a way to both reduce the greenhouse gas emissions from the farms and to slow growth in livestock production.

Another potential tool for slowing livestock production increases is a carbon tax. Even if a carbon tax is only imposed on fossil fuels, it would raise the price of synthetic nitrogen fertilizer and diesel fuel, and so indirectly increase the cost of meat, while reducing both CO_2 and N_2O emissions. Imposing a direct carbon tax on some foods runs into the challenge of figuring out the actual emissions associated with the production of those foods. An average emissions figure for a product such as beef hides the enormous variation between the emissions from a steer that spent its life on pasture managed to maximize soil carbon storage and received finishing supplements of food waste or grain raised in a cropping system managed for soil health and carbon, and a steer that started its life on degrading pasture or range and finished on feed raised without attention to soil health principles. Simply putting a carbon tax on a broad food category like beef, because on average beef has a high emissions intensity, could have the perverse effect of disproportionately harming the market for carefully managed low-emissions beef, because the combination of a tax and the added costs of managing for soil health and carbon would make the product too expensive. Even climate-conscious consumers might buy the cheaper higher emissions beef, assuming incorrectly that the tax accurately reflects and compensates for actual emissions.

Changing the management of publicly owned land offers another strategy to reduce overall livestock production. The Bureau of Land Management in the Department of Interior manages more than 10% of the total land area of the U.S. They lease roughly 60% of this land for grazing – more than 62 million hectares – primarily to white ranchers managing Eurasian breeds of cattle and sheep. Another 31 million hectares of U.S. Forest Service land are grazed by livestock, and other public agencies also lease land for grazing.[46] From an agroecological perspective it would make sense to transition these lands to raising native ruminants, including bison, antelope, and elk, rather than cattle and sheep, and to manage them

[45] CAFO stands for Concentrated Animal Feeding Operation and is defined as an operation with at least 1000 animal units confined on site for more than 45 days during the year. An animal unit is defined as 1000 pounds live weight and is typically calculated as 1000 beef animals, 700 dairy cows, 2500 pigs, 125,000 meat chickens, or 82,000 laying hens (NRCS, n.d.).

[46] Congressional Research Service (2017), U.S. Department of the Interior (2017), U.S. Government Accountability Office (2005).

primarily for habitat restoration, soil carbon storage, and indigenous food sovereignty.[47]

The question of how to manage grazing on these lands has been the subject of contention for over a century.[48] Federal agencies are required to consider "scientific, scenic, historical, ecological, environmental, air and atmospheric, water resource, and archeological values," as well as the economic interests of ranchers in the management of public lands, but critics have pointed out that the protection of these other values is uneven, depending both on the political orientation of the administration in power and on individual agency employees. Moreover, by law the fees for grazing on federal lands are set far below the market rate for private land and cover less than a quarter of the cost of administering the grazing program.[49] The ranchers who hold grazing rights on federal lands have disproportionate political influence, and have so far been able to prevent adjustments to fees and imposition of tighter management requirements. Nevertheless, given the urgency of mitigating climate change, the federal government has a responsibility to manage its landholdings in a way that stores soil carbon, protects the environment, and begins to redress the history of dispossessing first nations.

How much should we aim to reduce production of livestock? I don't have a specific recommendation; instead I propose some conceptual boundaries. At a minimum we want to raise and harvest enough livestock to support healthy natural ecosystems and sustainable cropping systems. As Chaps. 3 and 6 lay out, occasional grazing by ruminants is an integral component of sustainable cropping systems and healthy grassland ecosystems. On the other hand, there are strong economic pressures in the current food and agriculture system for farmers and ranchers to raise as many animals as they can without causing measurable water pollution or visible damage to the land. We don't have good information on how to determine the right number of animal units to maximize ecosystem health and soil carbon storage on a given site or for a region. We need research on that question in order to develop policy guidelines, as well as management recommendations for individual farms and ranches.

At the upper bound we should not be raising more livestock than people want to eat, or than is consistent with healthy nutrition. The current system where agribusiness seeks to persuade people to eat more meat and dairy than they want or need and where the food system promotes waste harms eaters and farmers as well as the climate.

[47] Thompson et al. (2020).

[48] Dana and Fairfax (1980).

[49] Davis (2008), Congressional Research Service (2017), Glaser et al. (2015), Vincent (2012).

9.7 Access to Food

Policies to reduce overproduction would likely improve income for farmers while reducing greenhouse gas emissions but would also put upward pressure on consumer prices for meat and dairy products. Even with cheap food prices, access to healthy food is an issue for more than 10% of the U.S. population. The question of how to bring livestock production into line with climate goals without further increasing disparities in food access has received surprisingly little attention. The reality is that meat and/or dairy are prized foods in most cultures, and also that the total amount of meat that can be produced with no net greenhouse gas emissions is less than the amount we are producing now with the help of fossil fuel subsidies in the form of synthetic nitrogen fertilizer, pesticides, and animal housing.[50] The reluctance so far of government to rein in overproduction reflects a legitimate concern about the social impact of rising food prices, as well as the political power of agribusiness.

However, although food supply can affect food security, it is not the only factor determining people's ability to eat well. The United Nations Food and Agriculture Organization lists four pillars of food security: physical availability of food, economic and physical access to food, food utilization, and stability.[51] Arguably the U.S. has focused too much on ensuring a large physical food supply and not enough on universal economic and physical access to food or economic stability. Low food prices alone do not guarantee economic or physical access. In the U.S., despite low food prices and high average income, more than 10% of households still experience occasional or chronic food insecurity because of income inequality and unequal physical access to healthy foods. In contrast, in southeast Asia, where food prices are relatively high, food security has increased substantially in recent decades as wages have risen.[52] In places like the U.S., where food prices are already low, providing direct financial assistance to lower income households would do more to increase food security than maintaining an excessively abundant food supply and very low prices. The income a carbon tax would generate could be used to fund such direct assistance, offsetting the impact of a slight increase in food prices.[53] During the COVID-19 pandemic of 2020 and 2021 the U.S. government was able to quickly provide money to the majority of U.S. adults under a certain income threshold, showing that this type of program is technically feasible.

Another possible approach to make the impacts of a carbon tax more equitable would be to make the tax progressive depending on use. This approach would be

---

[50] For example, a life cycle analysis comparing a multispecies pasture rotation to conventional livestock production in the southeastern U.S. found that the regenerative grazing had 66% lower greenhouse gas emissions per unit of meat than conventional production, but it also produced less than half as much meat per hectare as the conventional approach did (Rowntree et al. 2020).

[51] FAO (n.d.), Clay (2002), Commitee on World Food Security (2017).

[52] FAO (n.d.), Bauer (2020), Timmer (2000), Leroy et al. (2015), Coleman-Jensen et al. (2020).

[53] Marshall and Archer (2018).

technically more difficult to implement but might be more effective in changing consumption patterns, especially if the tax is visible. The idea would be that everyone would be entitled to some level of untaxed consumption of a food that has climate impacts, but that as an individual's consumption rises they should have to pay a progressively higher tax. For example, a person might be entitled to buy 5 kg of untaxed beef per year. When they go to buy the sixth kilo they would be taxed a certain percentage, and when they go to buy the tenth kilo the climate tax would go even higher. Having to pay the surcharge would be a reminder of the climate impacts of the purchase, as well as a financial disincentive.

9.8 Conclusion

Where are we in the challenge to (1) base policy on good science, (2) distribute the benefits and burdens of agricultural climate policy equitably and fairly, (3) monitor and enforce compliance with the policy, and (4) develop clearly articulated and shared goals? We still have work to do to get researchers, policy makers, and stakeholders to acknowledge and honestly negotiate around the messy, complex interactions of livestock and greenhouse gas fluxes. We have no system in place for monitoring livestock-related emissions, and we still face considerable technical challenges to measuring those emissions. Initial policy discussions have centered on voluntary incentives, leaving questions of compliance and free-ridership unanswered. Issues of equity and fairness increase the complexity of these common resource negotiations exponentially. Wealth, race, gender, education, ethnicity, and other factors intersect in their effect on individual equity, and economic and political power affect negotiations at the global level. There is increased acknowledgement of inequity, but the structures that maintain it are still strong and often unrecognized.

On the other hand, the U.S. has made amazing progress in recent months toward articulating a widely shared goal of reduced or even zero emissions for agriculture. This shared goal is creating tentative alliances among groups that have often been at odds, including agricultural and environmental groups at the national level, and my university extension colleagues in agriculture and natural resources closer to home. The ability of agroecosystems to store carbon in the soil means that it is possible to raise livestock with no net emissions, though we cannot raise enough to feed everyone the amount of meat and milk that Americans currently eat and throw away. Ideally, in 10 years I will be able to tell my neighbor that from a climate perspective he should simply eat the meat he likes best, because they are all raised with no net emissions.

The U.S. policy landscape around agriculture and climate change is changing at a dizzying speed right now. We seem to be moving toward a consensus goal of net zero greenhouse gas emissions, which is an amazing change from a few years ago. Private businesses, commodity groups, and farmer organizations are pledging to work toward reducing or eliminating agricultural greenhouse gas emissions, or even making agriculture a net sink for greenhouse gases. On April 20th, 2021 just before

Earth Day in the U.S., PepsiCo Inc. announced “a new, impact-driven Positive Agriculture ambition, anchored by a goal to spread regenerative farming practices across seven million acres, approximately equal to its entire agricultural footprint. The company estimates the effort will eliminate at least three million tons of greenhouse gas emissions (GHG) by the end of the decade.” Other major food companies like General Mills have made similar commitments, and representatives and senators in the U.S. Congress have proposed legislation to help accomplish these goals.[54]

These new developments are grounds for cautious optimism, but the swift shift in public discourse raises the fear that the policy pendulum could swing back just as quickly. The emphasis on voluntary approaches and the lack of enforcement mechanisms are also cause for concern. At the bottom of Pepsi’s press release announcing their new regenerative agriculture initiative they included this reservation:

> This release contains statements reflecting our views about our future performance that constitute “forward-looking statements” within the meaning of the Private Securities Litigation Reform Act of 1995. Forward-looking statements are generally identified through the inclusion of words such as “aim,” “anticipate,” “believe,” “drive,” “estimate,” “expect,” “goal,” “intend,” “may,” “plan,” “project,” “strategy,” “target” and “will” or similar statements or variations of such terms and other similar expressions. Forward-looking statements inherently involve risks and uncertainties that could cause actual results to differ materially from those predicted in such statements….[55]

PepsiCo is not unique in making no legally enforceable commitment to the admirable goals they publicize; they are only unusual in stating their lack of commitment where the public can see it. We cannot expect business or individuals to monitor the effectiveness of their own practices to reduce emissions or to enforce their rosy goals. Those actions are the responsibility of government and civil society.

At the same time, just as focusing exclusively on one narrow aspect of livestock greenhouse gas emissions does not create a foundation for good policy, focusing only on climate impacts is not enough. Agricultural policy must also address issues of equity, human health, animal welfare, biodiversity, and the many other challenges to making a good future for ourselves, our children, and our world. It is not an easy challenge, but it is one we can and must rise to.

Bibliography

Agricultural Marketing Service. (2016). *Removal of mandatory country of origin labeling requirements for beef and pork muscle cuts, ground beef, and ground pork.* Code of Federal Regulations. https://www.federalregister.gov/documents/2016/03/02/2016-04609/removal-of-mandatory-country-of-origin-labeling-requirements-for-beef-and-pork-muscle-cuts-ground

AGRICULTURE RESILIENCE ACT 117th Congress Section-by-Section Summary. (2021, April). National Sustainable Agriculture Coalition. https://sustainablea-

[54] PepsiCo (2021), General Mills (2021), “AGRICULTURE RESILIENCE ACT 117th Congress Section-by-Section Summary” (2021), Clayton (2021), U.S. Senate Committee on Agriculture, Nutrition, and Forestry (2021).

[55] PepsiCo (2021).

griculture.net/wp-content/uploads/2021/04/ARA-Section-by-Section-2021_FINAL.pdf?emci=58f93868-72a3-eb11-85aa-0050f237abef&emdi=ea000000-0000-0000-0000-000000000001&ceid=%7b%7bContactsEmailID%7d%7d

Bauer, L. (2020). *Hungry at thanksgiving: A fall 2020 update on food insecurity in the US* [Blog post]. Brookings Institute. https://www.brookings.edu/blog/up-front/2020/11/23/hungry-at-thanksgiving-a-fall-2020-update-on-food-insecurity-in-the-u-s/

Bloomberg Editorial Board. (2021, March 3). How farmers could fight climate change (and make a profit). *Bloomberg Opinion*. https://www.bloomberg.com/opinion/articles/2021-03-03/agricultural-carbon-markets-could-help-farmers-fight-climate-change

Brasher, P. (2021, April 20). Overhauled climate bill wins expanded support to lay groundwork for Ag carbon trading. *AgriPulse*. https://www.agri-pulse.com/articles/15720-new-carbon-market-bill-released

Brasher, P., Nuelle, B., & Wicks, N. (2021, March 3). Coronavirus food assistance program: Where did the money go? *AgriPulse*. https://www.agri-pulse.com/articles/15429-coronavirus-food-assistance-program-where-did-the-money-go

Brooks, R. (2021a, March 16). Carbon initiative incentive program is expanded for 2021. *AgWeb*. https://www.agweb.com/news/business/conservation/carbon-initiative-incentive-program-expanded-2021?mkt_tok=ODQzLVlHQi03OTMAAAF73qulz5nxQhlkykzIsD8uNkekAi1O8MtEOBiK-jmIaosFVP59mAvI7sTryYPzPtcyLDtLAAoG2UXCOjMjTeTDs6dILiE02C8qyshTQh0LSaXkwp8

Brooks, R. (2021b, April 19). Three Big Trends in the Carbon Marketplace. *AgWeb Farm Journal*. https://www.agweb.com/news/business/conservation/three-big-trends-carbon-marketplace

Bruckner, M., Häyhä, T., Giljum, S., Maus, V., Fischer, G., Tramberend, S., & Börner, J. (2019). Quantifying the global cropland footprint of the European Union's non-food bioeconomy. *Environmental Research Letters, 14*(4), 045011. https://doi.org/10.1088/1748-9326/ab07f5

Buehler, R., & Pucher, J. (2012). Cycling to work in 90 large American cities: New evidence on the role of bike paths and lanes. *Transportation, 39*(2), 409–432. https://doi.org/10.1007/s11116-011-9355-8

Callahan, S. (2019). *Land use, land value & tenure: Major land uses* [U.S. Department of Agriculture Economic Research Service]. https://www.ers.usda.gov/topics/farm-economy/land-use-land-value-tenure/major-land-uses/

Carpenter, S. (2012). The USDA discrimination cases: Pigford, in re black farmers, Keepseagle, Garcia, and love. *Drake Journal of Agricultural Law, 17*(1), 29.

Castagné, M., Lickel, S., Ritter, T., & Dufrasne, G. (2020). *Carbon markets and agriculture: Why offsetting is putting us on the wrong track*. Institute for Agriculture and Trade Policy. https://www.iatp.org/carbon-markets-and-agriculture

Clay, E. (2002). Food security: Concepts and measurement. In *Trade reforms and food security*. Food and Agriculture Organization of the United Nations. http://www.fao.org/3/y4671e/y4671e06.htm#fn21

Clayton, C. (2021, April 21). Groups support Ag Carbon Credit Bill: Growing climate solutions act returns with Bipartisan backing. *Progressive Farmer*. https://www.dtnpf.com/agriculture/web/ag/news/world-policy/article/2021/04/21/growing-climate-solutions-act-bibm

Coleman-Jensen, A., Rabbitt, M. P., Gregory, C. A., & Singh, A. (2020). *Household food security in the United States in 2019* (Economic research service no. 275; p. 47). U.S. Department of Agriculture, Economic Research Service. https://www.ers.usda.gov/webdocs/publications/99282/err-275.pdf?v=9623.4

Commitee on World Food Security. (2017). *Global strategic framework for food security and nutrition*. http://www.fao.org/3/MR173EN/mr173en.pdf

Congressional Research Service. (2017). *Statistics on livestock grazing on federal lands: FY2002 to FY2016* (p. 13). https://www.everycrsreport.com/files/20170828_R44932_72f6e4476960ca8f3ddf799ba21b9f3652ae11e2.pdf

Congressional Research Service. (2020). *Agricultural conservation: A guide to programs* (No. R40763). https://fas.org/sgp/crs/misc/R40763.pdf

Cubbage, S. (2021, April 14). Saving the planet is risky business. *Farm Journal AgWeb*. https://www.agweb.com/news/business/technology/saving-planet-risky-business?mkt_tok=ODQzLVlHQi03OTMAAAF8eSujhFuenKlcrN-3qENFIMvWh3EMB_m92j2sldt9Z54g-KXK31a9cJoup5B5Bap_aSIuYJDu5078sQBE0yD5s-bD4vzFZQwa0jUDmBKkc9CZ8ecQ

Cusack, D. F., Kazanski, C. E., Hedgpeth, A., Chow, K., Cordeiro, A. L., Karpman, J., & Ryals, R. (2021). Reducing climate impacts of beef production: A synthesis of life cycle assessments across management systems and global regions. *Global Change Biology, gcb.15509*. https://doi.org/10.1111/gcb.15509

Dana, S. T., & Fairfax, S. K. (1980). *Forest and range policy. Its development in the United States* (2nd ed.). McGraw-Hill Book Company.

Danley, B. (2019). Forest owner objectives typologies: Instruments for each owner type or instruments for most owner types? *Forest Policy and Economics, 105*, 72–82. https://doi.org/10.1016/j.forpol.2019.05.018

Davis, C. (2008). The politics of grazing on federal lands: A policy change perspective. In *Punctuated equilibrium and the dynamics of US environmental policy* (pp. 232–252). Yale University Press. https://doi-org.ezproxy.library.wisc.edu/10.12987/9780300129748-011

Dignac, M.-F., Derrien, D., Barré, P., Barot, S., Cécillon, L., Chenu, C., Chevallier, T., Freschet, G. T., Garnier, P., Guenet, B., Hedde, M., Klumpp, K., Lashermes, G., Maron, P.-A., Nunan, N., Roumet, C., & Basile-Doelsch, I. (2017). Increasing soil carbon storage: Mechanisms, effects of agricultural practices and proxies. A review. *Agronomy for Sustainable Development, 37*(2), 14. https://doi.org/10.1007/s13593-017-0421-2

FAO. (n.d.). *An introduction to the basic concepts of food security*. United Nations Food and Agriculture Organization. http://www.fao.org/3/al936e/al936e.pdf

Foley, J. (2021, June 17). *The world needs better climate pledges*. Project Drawdown. https://www.drawdown.org/news/insights/opinion-the-world-needs-better-climate-pledges

Food and Agriculture Climate Alliance. (2021). *Food and agriculture climate Alliance joint policy recommendations*. https://agclimatealliance.com/files/2020/11/faca_recommendations.pdf.

Food Forum, Food and Nutrition Board, Health and Medicine Division, & National Academies of Sciences, Engineering, and Medicine. (2019). *Sustainable diets, food, and nutrition: Proceedings of a workshop* (L. Pray, Ed.). National Academies Press. https://doi.org/10.17226/25192.

Fox Business. (2021, April). Biden's Green new deal means no meat for the 4th of July, have grilled Brussels sprouts instead. *Fox Business*https://www.foxbusiness.com/media/kudlow-bidens-climate plan-means-no-meat-for-the-4th-of-july-have-grilled-brussel-sprouts-instead

Garnett, T., Godde, C., Muller, A., Röös, E., Smith, P., & de Boer, I. (2017). *Ruminating on cattle, grazing systems, methane, nitrous oxide, the soil carbon sequestration question – and what it all means for greenhouse gas emissions*, p. 127.

General Mills. (2021). *Climate change*. General Mills. https://www.generalmills.com/en/Responsibility/Sustainability/climate-change

Gillon, S., Booth, E. G., & Rissman, A. R. (2016). Shifting drivers and static baselines in environmental governance: Challenges for improving and proving water quality outcomes. *Regional Environmental Change, 16*(3), 759–775. https://doi.org/10.1007/s10113-015-0787-0

Glaser, C., Romaniello, C., & Moskowitz, K. (2015). *CostsAndConsequences_01-2015.pdf*. Center for Biological Diversity. https://www.biologicaldiversity.org/programs/public_lands/grazing/pdfs/CostsAndConsequences_01-2015.pdf

Griffiths, C. (2021, March 1). The Carbon Contract Conundrum: Key considerations before signing up leased land. *Farm Journal AgWeb*. https://www.agweb.com/news/business/conservation/carbon-contract-conundrum?mkt_tok=eyJpIjoiTkRFellXVmpNR0ZrWkRaaylsInQiOiJtcVRmU1wvbVdGeFBtd0sxSmhXZWFZNGYyK2p3ZUhleWcxWFRwc09JamFNZWFZQVZJWHRmQ1UrajdMeVwvandtMVYxU1ZJNHM5WmdLenhOYVpTXC9Hb2hKb-kI1SXN3b2M3TWNuVVNCNW9PU2l0Nng0d0JNbVZwSVljM3NTV1I4RnluQiJ9

Gronewold, N. (2011, January 3). Chicago climate exchange closes Nation's first cap-and-trade system but keeps eye to the future. *New York Times*. https://archive.nytimes.com/www.nytimes.

com/cwire/2011/01/03/03climatewire-chicago-climate-exchange-closes-but-keeps-ey-78598.html?pagewanted=print

Harwatt, H., Sabaté, J., Eshel, G., Soret, S., & Ripple, W. (2017). Substituting beans for beef as a contribution toward US climate change targets. *Climatic Change, 143*(1–2), 261–270. https://doi.org/10.1007/s10584-017-1969-1

Hawken, P. (Ed.). (2017). *Drawdown: The most comprehensive plan ever proposed to reverse global warming*. Penguin. https://www.drawdown.org/solutions-summary-by-rank

Hendrickson, M. K. (2020). Covid lays bare the brittleness of a concentrated and consolidated food system. Agriculture and Human Values, 37(3), 579–580. https://doi.org/10.1007/s10460-020-10092-y

Klein, E. (2021, April 24). Let's launch a moonshot for meatless meat. *New York Times*. https://www.nytimes.com/2021/04/24/opinion/climate-change-meatless-meat.html

Kremen, C., & Merenlender, A. M. (2018). Landscapes that work for biodiversity and people. *Science, 362*(6412), eaau6020. https://doi.org/10.1126/science.aau6020

Laca, A.-L. (2020, December 9). *Dairy leaders praise USTR action on Canadian quota enforcement*. Dairy Herd Management. https://www.dairyherd.com/news/exports/dairy-leaders-praise-ustr-action-canadian-quota-enforcement

Lark, T. J., Salmon, J. M., & Gibbs, H. K. (2015). Cropland expansion outpaces agricultural and biofuel policies in the United States. *Environmental Research Letters, 10*(4), 044003. https://doi.org/10.1088/1748-9326/10/4/044003

Leonard, B., Parker, D. P., & Anderson, T. L. (2020). Land quality, land rights, and indigenous poverty. *Journal of Development Economics, 143*, 102435. https://doi.org/10.1016/j.jdeveco.2019.102435

Leroy, J. L., Ruel, M., Frongillo, E. A., Harris, J., & Ballard, T. J. (2015). Measuring the food access dimension of food security: A critical review and mapping of indicators. *Food and Nutrition Bulletin, 36*(2), 167–195. https://doi.org/10.1177/0379572115587274

Li, C., Frolking, S., & Butterbach-Bahl, K. (2005). Carbon sequestration in arable soils is likely to increase nitrous oxide emissions, offsetting reductions in climate radiative forcing. *Climatic Change, 72*(3), 321–338. https://doi.org/10.1007/s10584-005-6791-5

Lynch, J., Cain, M., Pierrehumbert, R., & Allen, M. (2020). Demonstrating GWP*: A means of reporting warming-equivalent emissions that captures the contrasting impacts of short- and long-lived climate pollutants. *Environmental Research Letters, 15*(4), 044023. https://doi.org/10.1088/1748-9326/ab6d7e

Lynch, J., Garnett, T., Persson, M., Röös, E., & Reisinger, A. (2020). *Methane and the sustainability of ruminant livestock* (p. 20). Food Climate Research Network.

Marshall, J., & Archer, R. (2018, August 31). *10 Fast facts about revenue-neutral carbon fees*. Citizens Climate Lobby. https://citizensclimatelobby.org/10-facts-revenue-neutral-carbon-fees/

Mayer, A. (2021, April 7). The down-low from DC: Ag Secretary Tom Vilsack on Climate Solutions. *AgWeb Farm Journal*. https://www.agweb.com/news/crops/crop-production/down-low-dc-ag-secretary-tom-vilsack-climate-solutions?mkt_tok=ODQzLVlHQi03OTMAAAF8T_fWdeOE2x_cnLoyoGa0UA6pCSmNKW3JRrWaJhrYHuxcK7dImKT6tQpbKRs0dCyIwoLS%2D%2DE6wdGUWuCetAnumzr2op9wseH2kh85eYuMSJ1AXK8

Mayerfeld, D. (2016). *Meat and our carbon footprint: Using deliberative dialogue to tackle contentious issues* (No. N4115). University of Wisconsin Extension. https://learningstore.extension.wisc.edu/products/meat-and-our-carbon-footprint-using-deliberative-dialogue-to-tackle-contentious-issues-p1804?_pos=1&_sid=1a0334302&_ss=r

McCammon, S. (2015, January 12). The story behind 40 acres and a mule. In *All things considered*. National Public Radio. https://www.npr.org/sections/codeswitch/2015/01/12/376781165/the-story-behind-40-acres-and-a-mule

McFetridge, S. (2021, April 7). *Biofuels producers, farmers not sold on switch to electric*. Washington Post. https://www.washingtonpost.com/business/technology/biofuels-producers-farmers-not-sold-on-switch-to-electric/2021/04/07/b581688c-97b3-11eb-8f0a-3384cf4fb399_story.html

Mitchell, W. (2005). Destabilizing the normalization of rural black land loss: A critical role for legal empiricism. *Wisconsin Law Review, 2*, 557–616.

Montagnini, F., & Nair, P. K. R. (2004). Carbon sequestration: An underexploited environmental benefit of agroforestry systems. *Agroforestry Systems, 61*(1–3), 281–295.

Morgan, T. (2021, April 8). Less than 1% of farmers have entered into a carbon contract, survey shows. *AgWeb Farm Journal.* https://www.agweb.com/news/business/conservation/less-1-farmers-have-entered-carbon-contract-survey-shows

National Sustainable Agriculture Coalition. (2019). *Agriculture and climate change: Policy imperatives and opportunities to help producers meet the challenge.* https://sustainableagriculture.net/wp-content/uploads/2019/11/NSAC-Climate-Change-Policy-Position_paper-112019_WEB.pdf

Newkirk, V. R. I. (2019, September). The great land robbery: The shameful story of how 1 million black families have been ripped from their farms. *The Atlantic.* https://www.theatlantic.com/magazine/archive/2019/09/this-land-was-our-land/594742/

Newman, L. (2020). The promise and peril of "Cultured Meat". In R. Katz-Rosene & S. J. Martin (Eds.), *Green meat? Sustaining eaters, animals, and the planet.* McGill-Queen's University Press.

Nickerson, C., & Borchers, A. (2012, March 1). How is land in the United States used? A focus on agricultural land. *Amber Waves.* https://www.ers.usda.gov/amber-waves/2012/march/data-feature-how-is-land-used/

NRCS. (2021a). *NRCS conservation programs: Conservation stewardship program (CSP).* US Department of Agriculture, Natural Resource Conservation Service. https://www.nrcs.usda.gov/Internet/NRCS_RCA/reports/fb08_cp_cstp.html

NRCS. (2021b). *NRCS conservation programs: Environmental quality incentives program (EQIP).* US Department of Agriculture, Natural Resource Conservation Service. https://www.nrcs.usda.gov/Internet/NRCS_RCA/reports/fb08_cp_eqip.html

NRCS. (n.d.). *Animal feeding operations* [US Department of Agriculture]. Natural Resources Conservation Service. https://www.nrcs.usda.gov/wps/portal/nrcs/main/national/plantsanimals/livestock/afo/

Ocasio-Cortez, A. (2019). *Recognizing the duty of the Federal Government to create a Green new Deal.* US Government Printing Office. https://www.congress.gov/116/bills/hres109/BILLS-116hres109ih.pdf

Osmond, D. L., Hoag, D. L. K., Luloff, A. E., Meals, D. W., & Neas, K. (2015). Farmers' use of nutrient management: Lessons from watershed case studies. *Journal of Environmental Quality, 44*(2), 382–390. https://doi.org/10.2134/jeq2014.02.0091

Ostrom, E., Burger, J., Field, C. B., Norgaard, R. B., & Policansky, D. (1999). Revisiting the commons: Local lessons. *Global Challenges, 284*, 6.

Ostrom, E. E., Dietz, T. E., Dolšak, N. E., Stern, P. C., Stonich, S. E., & Weber, E. U. (Eds.). (2002). *The drama of the commons.* National Academy Press.

Patel-Weynand, T., Bentrup, G., & Schoeneberger, M. M. (2018). Agroforestry: Enhancing resiliency in U.S. In *Agricultural landscapes under changing conditions: Executive summary* (WO-GTR-96a). U.S. Department of Agriculture, Forest Service. https://doi.org/10.2737/WO-GTR-96a

PepsiCo. (2021, April 20). *PepsiCo announces 2030 goal to scale regenerative farming practices across 7 million acres, equivalent to entire agricultural footprint.* PepsiCo. https://www.pepsico.com/news/press-release/pepsico-announces-2030-goal-to-scale-regenerative-farming-practices-across-7-mil04202021

Poore, J., & Nemecek, T. (2018). Reducing food's environmental impacts through producers and consumers. *Science, 360*(6392), 987–992. https://doi.org/10.1126/science.aaq0216

Porter, P. A., Mitchell, R. B., & Moore, K. J. (2015). Reducing hypoxia in the Gulf of Mexico: Reimagining a more resilient agricultural landscape in the Mississippi River watershed. *Journal of Soil and Water Conservation, 70*(3), 63A–68A. https://doi.org/10.2489/jswc.70.3.63A

Pucher, J., & Buehler, R. (2008). Making cycling irresistible: Lessons from the Netherlands, Denmark and Germany. *Transport Reviews, 28*(4), 495–528. https://doi.org/10.1080/01441640701806612
Richmond, T. (2021, May 2). *White farmers sue seeking government loan forgiveness*. Associated Press. https://www.morningagclips.com/white-farmers-sue-seeking-government-loan-forgiveness/
Rippon-Butler, H. (2020). *Land policy: Towards a more equitable farming future*. National Young Farmers Coalition. https://www.youngfarmers.org/land/wp-content/uploads/2020/11/LandPolicyReport.pdf?emci=061d8bd0-259e-eb11-85aa-0050f237abef&emdi=202a89dc-ff9e-eb11-85aa-0050f237abef&ceid=1994345
Rissman, A. R., Kohl, P. A., & Wardropper, C. B. (2017). Public support for carrot, stick, and no-government water quality policies. *Environmental Science & Policy, 76*, 82–89. https://doi.org/10.1016/j.envsci.2017.04.012
Rowntree, J. E., Stanley, P. L., Maciel, I. C. F., Thorbecke, M., Rosenzweig, S. T., Hancock, D. W., Guzman, A., & Raven, M. R. (2020). Ecosystem impacts and productive capacity of a multi-species pastured livestock system. *Frontiers in Sustainable Food Systems, 4*, 13.
Saitone, T., Schaefer, K. A., & Scheitrum, D. P. (2021). COVID-19 morbidity and mortality in U.S. meatpacking counties. *Food Policy, 102072*. https://doi.org/10.1016/j.foodpol.2021.102072
Sanburg, H. (2021, February 23). A World Without The Beef Checkoff. *Drovers*. https://www.drovers.com/news/industry/sanburg-world-without-beef-checkoff
Sanford, G. R., Posner, J. L., Jackson, R. D., Kucharik, C. J., Hedtcke, J. L., & Lin, T.-L. (2012). Soil carbon lost from Mollisols of the North Central U.S.A. with 20 years of agricultural best management practices. *Agriculture, Ecosystems & Environment, 162*, 68–76. https://doi.org/10.1016/j.agee.2012.08.011
Schlesinger, W. H., & Amundson, R. (2019). Managing for soil carbon sequestration: Let's get realistic. *Global Change Biology, 25*(2), 386–389. https://doi.org/10.1111/gcb.14478
Schulte, L. A., Niemi, J., Helmers, M., Liebman, M., Arbuckle, J. G., James, D. E., Kolkag, R. K., O'Nealh, M. E., Tomerf, M. D., Tyndalla, J. C., Asbjornseni, H., Drobneyj, P., Nealk, J., Van Ryswykl, G., & Wittec, C. (2017). Prairie strips improve biodiversity and the delivery of multiple ecosystem services from corn–soybean croplands. *Proceedings of the National Academy of Sciences, 114*(50), E10851. https://doi.org/10.1073/pnas.1719680114
Stanley, P. L., Rowntree, J. E., Beede, D. K., DeLonge, M. S., & Hamm, M. W. (2018). Impacts of soil carbon sequestration on life cycle greenhouse gas emissions in Midwestern USA beef finishing systems. *Agricultural Systems, 162*, 249–258. https://doi.org/10.1016/j.agsy.2018.02.003
Taylor, C. A., Boulos, C., & Almond, D. (2020). Livestock plants and COVID-19 transmission. *Proceedings of the National Academy of Sciences, 202010115*. https://doi.org/10.1073/pnas.2010115117
Thompson, S., Pritty, P., & Thapa, K. (2020). Eco-Carnivorism in Garden Hill first nation. In *Green meat?: Sustaining eaters animals and the planet* (pp. 107–132). McGill-Queen's University Press.
Timmer, C. P. (2000). The macro dimensions of food security: Economic growth, equitable distribution, and food price stability. *Food Policy, 25*(3), 283–295. https://doi.org/10.1016/S0306-9192(00)00007-5
Toensmeier, E. (2016). The carbon farming solution: A global toolkit of perennial crops and regenerative agriculture practices for climate change mitigation and food security. .
Tuomisto, H. L., & Teixeira de Mattos, M. J. (2011). Environmental impacts of cultured meat production. *Environmental Science & Technology, 45*(14), 6117–6123. https://doi.org/10.1021/es200130u
U.S. Department of Agriculture. (2021a). *Notice of request for public comment on the executive order on tackling the climate crisis at home and abroad*. Federal Register. https://www.federalregister.gov/documents/2021/03/16/2021-05287/notice-of-request-for-public-comment-on-the-executive-order-on-tackling-the-climate-crisis-at-home
U.S. Department of Agriculture. (2021b). *After Identifying Gaps in Previous Aid, USDA Announces 'Pandemic Assistance for Producers' to Distribute Resources More Equitably*.

https://www.usda.gov/media/press-releases/2021/03/24/after-identifying-gaps-previous-aid-usda-announces-pandemic

U.S. Department of Agriculture. (n.d.-a). *Soil Health.* Farmers.Gov. Retrieved August 4, 2021, from https://www.farmers.gov/conservation/soil-health

U.S. Department of Agriculture. (n.d.-b). *Country of origin Labeling (COOL).* Agricultural Marketing Service. https://www.ams.usda.gov/rules-regulations/cool

U.S. Department of Agriculture, & U.S. Department of Health and Human Services. (2020). Dietary guidelines for Americans 2020–2025. U.S. Department of Agriculture. https://dietaryguidelines.gov/sites/default/files/2021-03/Dietary_Guidelines_for_Americans-2020-2025.pdf

U.S. Department of Agriculture Agricultural Marketing Service. (n.d.). *Research & promotion programs.* Agricultural Marketing Service. Retrieved May 5, 2021, from https://www.ams.usda.gov/rules-regulations/research-promotion

U.S. Department of Agriculture Economic Research Service. (2021). *Feed grains database. Feed Grains Database Feed Grains Custom Query.* https://data.ers.usda.gov/FEED-GRAINS-custom-query.aspx

U.S. Department of the Interior. (2017). *8 things you didn't know about the bureau of land management.* https://www.doi.gov/blog/8-things-you-didnt-know-about-bureau-land-management

U.S. Environmental Protection Agency. (n.d.-a). Clean water act Section 404 and agriculture. *Section 404 of the Clean Water Act.* Retrieved September 3, 2021, from https://www.epa.gov/cwa-404/clean-water-act-section-404-and-agriculture

U.S. Environmental Protection Agency. (n.d.-b). *Livestock anaerobic digester database.* Retrieved March 11, 2021, from https://www.epa.gov/agstar/livestock-anaerobic-digester-database

U.S. Government Accountability Office. (2005). *Livestock grazing: Federal expenditures and receipts vary, depending on the agency and the purpose of the fee charged.* https://www.govinfo.gov/content/pkg/GAOREPORTS-GAO-05-869/html/GAOREPORTS-GAO-05-869.htm

USDA. (n.d.). *North central sustainable agriculture research and education.* Retrieved May 3, 2021, from https://northcentral.sare.org/grants/apply-for-a-grant/research-and-education-grant/

Union of Concerned Scientists. (2020). Each country's share of CO2 emissions. https://www.ucsusa.org/resources/each-countrys-share-co2-emissions

Upadhaya, S., Arbuckle, J. G., & Schulte, L. A. (2021). Developing farmer typologies to inform conservation outreach in agricultural landscapes. *Land Use Policy, 101*, 105157. https://doi.org/10.1016/j.landusepol.2020.105157

US Senate Committee on Agriculture, Nutrition, and Forestry. (2021, April 22). In case you missed it: Here's What They're Saying About the Growing Climate Solutions Act. https://www.agriculture.senate.gov/newsroom/dem/press/release/in-case-you-missed-it-heres-what-theyre-saying-about-the-growing-climate-solutions-act.

Vincent, C. H. (2012). *Grazing fees: Overview and issues* (Congressional Research Service No. RS21232). https://www.emwh.org/pdf/conservation/forage/Grazing%20Fees%20Overview%20and%20Issues.pdf

Voisin, A. (1959). *Grass Productivity* (Herriot, Trans.; reprinted 1988). Island Press.

Wilson, J. (2002). Scientific uncertainty, complex systems, and the design of common-pool institutions. In *The drama of the commons.* National Academy Press.

Wisconsin Initiative on Climate Change Impacts. (2020). *Report to the Governor's task force on climate change.*https://wicci.wisc.edu/wp-content/uploads/wicci-report-to-governors-task-force.pdf

Wisconsin Initiative on Climate Change Impacts. (2022). *Wisconsin's changing climate: Impacts and solutions for a warmer climate.* Nelson Institute for Environmental Studies, University of Wisconsin-Madison, and the Wisconsin Department of Natural Resources. https://wicci.wisc.edu/2021-assessment-report/full-report/

World Bank. (2018). *Agricultural land (% of land area).*https://data.worldbank.org/indicator/AG.LND.AGRI.ZS

Index

D. Mayerfeld (ed.), *Our Carbon Hoofprint*, Food and Health,
https://doi.org/10.1007/978-3-031-09023-3

MIX
Papier aus verantwortungsvollen Quellen
Paper from responsible sources
FSC® C105338

If you have any concerns about our products,
you can contact us on
ProductSafety@springernature.com

In case Publisher is established outside the EU,
the EU authorized representative is:
Springer Nature Customer Service Center GmbH
Europaplatz 3, 69115 Heidelberg, Germany

Printed by Libri Plureos GmbH
in Hamburg, Germany